# DECISION MAKING

A Psychophysics Application of Network Science

**STUDIES OF NONLINEAR PHENOMENA IN LIFE SCIENCE***

*Editor-in-Charge:* Bruce J. West

---

Vol. 4 Physical Theory in Biology — Foundations and Explorations
*edited by C J Lumsden, W A Brandts & L E H Trainor*

Vol. 5 Nonlinear Dynamics in Human Behavior
*edited by W Sulis & A Combs*

Vol. 6 The Complex Matters of the Mind
*edited by F Orsucci*

Vol. 7 Physiology, Promiscuity, and Prophecy at the Millennium: A Tale of Tails
*by B J West*

Vol. 8 Dynamics, Synergetics, Autonomous Agents: Nonlinear Systems Approaches to Cognitive Psychology and Cognitive Science
*edited by W Tschacher & J-P Dauwalder*

Vol. 9 Changing Mind: Transitions in Natural and Artificial Environments
*by F F Orsucci*

Vol. 10 The Dynamical Systems Approach to Cognition: Concepts and Empirical Paradigms based on Self-Organization, Embodiment, and Coordination Dynamics
*edited by W Tschacher & J-P Dauwalder*

Vol. 11 Where Medicine Went Wrong: Rediscovering the Path to Complexity
*by B J West*

Vol. 12 Mind Force: On Human Attractions
*by F Orsucci*

Vol. 13 Disrupted Networks: From Physics to Climate Change
*by B J West & N Scafetta*

Vol. 14 Fractal Time: Why a Watched Kettle Never Boils
*by S Vrobel*

Vol. 15 Decision Making: A Psychophysics Application of Network Science
*edited by P Grigolini & B J West*

*For the complete list of titles in this series, please go to
http://www.worldscibooks.com/series/snpls_series

Studies of Nonlinear Phenomena in Life Science – Vol. 15

# DECISION MAKING

## A Psychophysics Application of Network Science

Center for Nonlinear Science, University of North Texas, USA
10 – 13 January 2010

*editors*

**Paolo Grigolini**
*University of North Texas, USA*

**Bruce J West**
*Army Research Office, USA*

World Scientific

NEW JERSEY • LONDON • SINGAPORE • BEIJING • SHANGHAI • HONG KONG • TAIPEI • CHENNAI

*Published by*

World Scientific Publishing Co. Pte. Ltd.
5 Toh Tuck Link, Singapore 596224
*USA office:* 27 Warren Street, Suite 401-402, Hackensack, NJ 07601
*UK office:* 57 Shelton Street, Covent Garden, London WC2H 9HE

**British Library Cataloguing-in-Publication Data**
A catalogue record for this book is available from the British Library.

**Studies of Nonlinear Phenomena in Life Science — Vol. 15**
**DECISION MAKING**
**A Psychophysics Application of Network Science**



ISBN-13 978-981-4365-81-9
ISBN-10 981-4365-81-5

Printed in Singapore by Mainland Press Pte Ltd.

# PREFACE

The Center for Nonlinear Science (CNS) of the University of North Texas hosted the workshop *Decision Making, A Psychophysics Application of Network Science* in the second week of January 2010. The workshop brought together a select group of leading international scientists with Army scientists to address and review the scientific issues associated with the networking of decision making. The workshop constituted a review and critique of the second year of the Army Research Office/Army Research Laboratory (ARO/ARL) project: *Network Theory of Human Decision Making*. Other Army scientists participating in the ARL research program on Network Science participated in the workshop by means of video conferencing.

The workshop format was based on the interaction among four areas of specialization of at least four participants each, for a total of 16 invited speakers. The members of the workshop who did not present participated through question and answer sessions and informal discussion. The four categories are:

(1) Psychological testing and modeling: Decision making under stress has been proven to generate a transition from 1/f to white noise in cognitive fluctuations. The CNS group has proposed a model, based on the Weber–Fechner law in perception theory, to explain this effect: one of the focuses of the workshop was to discuss how to relate this and other phenomenological models to the dynamics of a network of interacting neurons.

(2) Neurophysiology and the brain dynamics: The CNS group developed a dynamic approach to theoretical predictions of the phenomenological model in (1) by means of a network of interacting neurons. The spectra of brain activity and the state of the art of dynamically modeling that activity were discussed from the perspective of professional neurophysiologists.

(3) Information transport: A new theory was proposed for the information transport in complex networks, called the Principle of Complexity Management (PCM), which establishes that the transport of information from a stimulus to a complex network

# CONTENTS

# OVERVIEW OF ARO PROGRAM ON NETWORK SCIENCE FOR HUMAN DECISION MAKING

BRUCE J. WEST
*Information Science Directorate,*
*US Army Research Office,*
*Durham, NC, 27709*

This program brings together researchers from disparate disciplines to work on a complex research problem that defies confinement within any single discipline. Consequently, not only are new and rewarding solutions sought and obtained for a problem of importance to society and the Army, that is, the human dimension of complex networks, but, in addition, collaborations are established that would not otherwise have formed given the traditional disciplinary compartmentalization of research. This program develops the basic research foundation of a science of networks supporting the linkage between the physical and human (cognitive and social) domains as they relate to human decision making. The strategy is to extend the recent methods of non-equilibrium statistical physics to non-stationary, renewal stochastic processes that appear to be characteristic of the interactions among nodes in complex networks. We also pursue understanding of the phenomenon of synchronization, whose mathematical formulation has recently provided insight into how complex networks reach accommodation and cooperation. The theoretical analyses of complex networks, although mathematically rigorous, often elude analytic solutions and require computer simulation and computation to analyze the underlying dynamic process.

## 1. Introduction

Modern society is more interconnected than humanity has been at any time in world history. A western city could not function without garbage collection, interconnected sewers, electricity from the power grid, transportation networks, food distribution networks, health care networks and would have a very different form if it did not have networks of education, banking, telephone service and the Internet. Some of these support networks within the city are physical, others are social and their forms have been evolving for millennia. Part of that evolution was the development of their interoperability such that these networks are all interconnected and in one way or another they connect to national and/or global networks [1]. This network-of-networks (NoN) is the engineered webbing of humanity, but there are comparable structures in the biosphere and ecosphere involving plant and animal NoN of unlimited variety. Clues to understanding human-based complex adaptive networks* (CAN) may be found in naturally occurring CANs.

This modernity is manifest in the military through the development of Network-Centric Warfare (NCW) which takes cognizance of human behavior in a networked environment of organized actions, often violent, directed toward political, social and military ends and is the basis of a new theory of war [2,3]. Thus, NCW has at its core a shift in focus from the platform to the network of which the platform is a member. Army scientists need to understand the dynamics, controllability and predictability of generic nonlinear complex networks in order to realize their goal of supporting both society and the soldier through research and the development of new technologies.

It is not only our external world that is cluttered with networks, but our internal world as well. The neuronal network carrying the brain's signal to the body's physiological networks is even more complex than the modern city or typical ecological network. Consequently it is premature to assign preeminence to any one CAN or collection of CANs from any particular discipline before we understand the patterns formed by the multiple interactions within and among them. Thus, the basic research into network science must span and encompass a multitude of disciplines.

The problem we address within this program is to develop the basic research foundation of a science of networks that supports the linkage between the physical and human (cognitive and social) domains as they relate to human decision-making. This approach is not directed at the totality of developing a Network Science, but has the more modest goal

---

*Note that this definition is related to, but is distinct from, complex adaptive systems (CAS) introduced in the literature on complex dynamical systems. This is different from the more traditional use of computers, which is to 'merely' evaluate theoretical predictions and process data.

of understanding the deeply interdependent physical and human networks of crucial importance to society as a whole and to the Army in particular. Even such a restricted problem is a significant challenge due to the multiply interconnecting networks buttressing the common decision making objective; the research is uncovering this interconnectedness of the human and physical CANs, suggesting how to control them and make predictions. Achieving this goal requires collaboration of scientists from multiple disciplines, and the program has provided a context for the Army Research Laboratory (ARL) scientists in the Computer and Information Science Directorate (CISD); the Human Research Engineering Directorate (HRED); the Army Research Office (ARO), as well as other ARL scientists, to focus on the interactions of the human, information and communication networks.

On the one hand, the military is proactive in that networked forces can operate in an agile manner to promote decision-making superiority. On the other hand, it is reactive in the need to respond to enemies who are also using the power of networks against United States interests. The research program provides insight to allow the Army to anticipate the enemy's use of network strategy and thereby reduce the reactive mode of operation. Consequently, we proceed by identifying what is presently known about complex networks, regardless of the disciplinary context and adapt that understanding to the decision-making paradigm. Moreover, the barriers to further understanding and the gaps in knowledge of the linkages between physical and human decision-making networks are addressed.

The research strategy of this program is based on theory, computation/simulation and experiment/observation. This is a cyclic interactive process in which new theory stimulates unique simulations, yielding insight into parameter values and network configurations, which in turn suggests specific experiments, whose outcome guides the refinement and development of theory. This modern approach to scientific research[†] is applied to the phenomenon of human decision-making in NCW. The core group of Army scientists is the focal point for

[†]Note that what makes this approach modern is the inclusion of computation/simulation in the iterative scientific process that historically included only theory and experiment.

external researchers requiring militarily relevant challenges and internal ARL efforts.

## 2. Background

The pace of growth of information and communication technology, along with network capabilities is driving fundamental conceptual changes in the United States and coalition defense doctrine. Three of the critical drivers of this change are the information, communication and social networks, which have worked symbiotically to foster the changes in the nature of military operations. Certainly these three types of networks taken separately do not constitute new phenomena. Twenty five years ago social networks were fairly stable, being maintained by communication over telephone and mailing networks, and supported by libraries and other information archives. Today, social network are ephemeral, with virtual interactions in cyberspace chat rooms, information being supplied by Wikipedia and all available through Google. Social networks are large, ad hoc, and maintained in real time by the Internet services and cell phone technology. While in previous decades it would have been unthinkable for the average soldier to circumvent the chain of command and address his/her concerns directly to a general officer, the modern day computer with the network of emails actually invites such interactions.

The research program promotes basic research investigations into how these networks in the greater society are driving changes in the Army's information, communication and social domains and how these changes, in turn, influence and adapt decision-making.

### 2.1. *What we know about networks*

To frame the research conducted within this program we briefly review the limited knowledge of the basic properties of certain complex networks, starting with the simplest, in which the connections among the nodes (the fundamental elements of the network) are random. Random networks are static and although they do demonstrate a number of interesting properties, they are too restrictive to mimic the full range of behaviors of most real-world phenomena. Random networks are characterized by average quantities. Two real-world properties that random networks do *not* capture are growth and the criteria for establishing new connections in growing networks. *Preferential attachment* is a rule by which newly formed nodes prefer to make connections to nodes having the greatest number of existing connections within a network. This rule, along with growth, is sufficient to develop an understanding of a class of scale-free networks. Such scale-free networks have inverse power-law distributions in the number of connections to a given node. There is a deep relationship between these scale-free networks and fractal statistics, for a more complete discussion see, for example, West [4].

The question arises as to why scaling networks appear in such a variety of contexts; biological, chemical, economic, social, physical and physiological. Answers to this question have been found independently by a number of investigators and are all of a similar nature. In the case of the human lung West [5] determined that a physiological fractal network was pre-adapted to errors, is unresponsive to random perturbations and therefore has a decided evolutionary advantage. The evolutionary advantage would lead one to expect to find inverse power laws in all manner of biological phenomena, which we do. A decade later a similar result was found for scaling networks, for example, the Internet was shown to be fairly unresponsive to either random attacks or random failures [6].

An inverse power law in the number of links implies a large number of nodes with relatively few connections, but a non-negligible number of nodes could be considered to be *hubs*. Consequently, random attacks on such networks would more likely destroy or disrupt nodes with only a few connections and therefore have negligible effect. The likelihood of striking a hub at random and thereby having a global effect is relatively small. Therefore the scale-free network is robust against random attacks. It is this tolerance to local attack (failure) that gives rise to the robustness of scale-free networks and this tolerance is a consequence of the architecture of the network.

The scale-free network and fractal physiological structures both have an evolutionary advantage because of their fractal properties. On the other hand, such networks are susceptible to attacks directed at hubs, and elimination of only one or two of these hubs could produce catastrophic failure of the network. It is this later strategy that is adopted by terrorists, when these most sensitive elements of the network can be identified and attacked [6].

The scale-free network formalism was constructed using a combination of data analysis, reasonable arguments concerning the interactions among individuals in a social network and mathematical modeling. The distribution of network connections resulting from this process was inverse power law in agreement with observation. A completely different approach was taken by Carlson and Doyle [7], who asked what kind of control process, was necessary to design a network having inverse power-law behavior. This is an engineering approach in which the constraints on the design of the network must be specified. Although both approaches result in inverse power-law distributions in agreement with data, other properties of the network are remarkably different.

The HOT[‡] procedure is intended to provide robust network performance in the face of environmental uncertainties and replaces scale-free with scale-rich networks. This engineering perspective forces power laws in specified network properties through tradeoffs between yield, cost of resources and tolerance of risks. The result is a "robust, yet fragile" network behavior. The difference between scale-free and scale-rich is that in the former large fluctuations emerge and recede as a natural consequence of the internal dynamics. Consequently the network is robust. On the other hand, the latter network is hypersensitive to new perturbations that were not part of the design; so that the HOT network is robust, yet fragile. It should be emphasized that both types of networks have their Achilles' heels. How to identify these weaknesses and design around them is one of the remaining challenges.

It has been determined that inverse power laws are ubiquitous in complex networks, appearing in every context from the smallest molecular level to the largest social gathering. Moreover the existence of

---

[‡]The acronym HOT, originally short for *Highly Optimized Tolerance* has subsequently been used to mean *Heavily Organized Tradeoffs* and *Heuristically Optimized Tolerance*, and now it is used for any of a number of interchanges of the three words.

such inverse power-law statistics emphasizes the differences between simple networks that can be described by the statistics of Gauss and the complex networks which cannot. However, knowing the statistics of a network to be inverse power law is <u>not</u> sufficient to understand the underlying generating mechanisms. This is a question that has been answered in part in this program and is discussed by West and Grigolini [8]. We have identified the mathematical problems whose solutions contribute in a substantial way to the development of the measures required to separate and categorized inverse power laws according to their dynamics.

### 2.2. *What we do not know about the linking of physical and human networks*

A number of general research challenges in Network Science were set out in the National Research Council report in 2005 [1], based on a survey of academic institutions from which they learned;

> *...developing predictive models of the behavior of large, complex networks is difficult. There are relatively few rigorous results to describe the scaling of their behaviors with increasing size. Surprisingly, this is true for common engineered networks like the Internet as well as for social and biological networks.*

Herein we modify the committee's broadly based research challenges and realign them with the purposes of the physical-human networks addressed in this proposal.

- *Dynamics, spatial location, and information propagation in networks.* The structure-function relation long identified as a fundamental problem in biology defines one of the basic challenges for networks arising in the physical and human domains. For example the statistics of information packets on complex networks do not have Poisson statistics but are clustered in time, occurring in bursts, requiring distributions with heavy tails. But the dynamics of such distributions are only partially understood and require further development [9].

- *Modeling and analysis of very large networks.* Tools, abstractions, and approximations do not exist that allow reasoning about large-scale networks. Techniques that have modeled noisy and incomplete data in physical networks need to be extended to human networks.
- *Design and synthesis of networks.* With one or two exceptions, techniques are not available to design or modify a network to obtain desired properties of complex dynamic networks, particularly when multiple conflicting constraints must be satisfied as in CANs or NoN.
- *Increasing the level of rigor and mathematical structure.* A survey of the university campuses in the United States revealed a consensus that the state of the art in network science does not have an appropriately rigorous mathematical basis.
- *Abstracting the common concepts across fields.* A set of common concepts for network science does not exist that is defined across the various scientific disciplines. There is discipline-specific nomenclature that has been developed for specialized needs, but the equivalence of these terminologies for a discipline-independent characterization of networks has not been established.
- *Better experiments and measurements of network structure.* There is only a limited set of tools available for investigating the structure and function of large-scale networks, in part, because of the paucity of current data sets.
- *Robustness and security of networks.* There is a lack of understanding of networked systems and no design methodology that is both robust to variations in the components and secure against hostile intent.
- *Decision making in an information rich environment.* There is a lack of understanding in the relationship between the network structures and complexity, and the impact of this on organizational design and individual/unit behaviors.

## 3. What We Have Been Doing

To understand a networked environment we must recognize the complex relationships within and among communications and information technologies, as well as the corresponding communications, information and social networks. These networks can be thought of as CANs. Understanding the behavior of a CAN is important not only for the military, but also to understand the nature of asymmetric threats. Warfare must be analyzed as a complex network that is linked and interactive with the surrounding social-economic and political context. Complex networks share four properties: emergence, nonlinearity, uncertainty and adaptation. Essentially, these networks grow and evolve in unanticipated ways, they operate in a unique fashion that is not orderly, changing in spurts or in small numbers and they adapt to their surroundings and to their own changing behavior. ARL scientists must understand the interplay of these four features of complex networks to develop NCW behavioral and organizational decision-making strategies.

Though NCW is dependent upon physical, information and communication networks, it is fundamentally about human and organizational behavior, and requires a new way of thinking about military operations [10]. The Army investment in network research in the cognitive and social domains of decision-making in an information-rich environment has been limited. This gap is addressed by the research in the present program.

One of the mysteries of human social interaction is how agreements are reached and cooperative alliances are made. At the end of the nineteenth century the separation of phenomena into simple and complex was relatively straightforward. A simple network was one that could be described by one or a few variables, and whose equations of motion could be given, for example, by Newton's laws. In such networks the initial conditions are specified and the final state is calculated (predicted) by solving the equations of motion. The predicted behavior of the network, typically a particle trajectory, is then compared with the result of experiment and if the two agree, within a pre-established degree of accuracy, the simple model provides a faithful description of the phenomenon. Thus, simple physical networks have simple descriptions.

Note that we do not discuss nonlinear dynamics and chaos here, since these concepts did not influence the nineteenth century scientists applying physics to the understanding of society. However, we find that the modern understanding of complex networks does require the use of these techniques.

As more elements are added to a network there are more and more interactions, and the relative importance of any single interaction diminishes proportionately. There comes a point at which the properties of the network are no longer determined by the individual trajectories, but only by the averages over a large number of such trajectories. This is how the statistical picture of a phenomenon replaced the individual trajectory description. The erratic single trajectory is replaced with a distribution function that describes an ensemble of trajectories and the equations of motion for individual elements are replaced with an equation of motion for the probability distribution function.

Individuals become part of the 'mob' or social group in a number of ways: choice, peer pressure and subliminal seduction; but always through a sequence of decisions, either conscious or not. The mathematical modeling of cooperative behavior between and among CANs is often found in the context of synchronization [11] frequently seen as the manner in which order emerges from chaos. Biological oscillators such as pacemaker cells in the human heart are notoriously poor clocks, which is to say, that the firing period varies from cell to cell and from firing to firing. However, collections of such naturally variable entities when allowed to interact become phase locked and lose their individual variability. How natural phenomena accomplish this phase synchronization has been gaining increasing attention since Turing's early work on the modeling of collective phenomena and informs the discussion on how social organizations achieve consensus, which we discuss in the context of our own work elsewhere in this volume [8]. The more recent mathematical investigations into the coupling mechanisms of such collectives was made by Kuramoto [12] using a network of coupled nonlinear oscillators to mimic natural synchronization in oscillating chemical reactions. A similar strategy was employed by Winfree [13] in his study of cardiac oscillations.

Multiple-scales and across-scale coupling are recurrent themes in the study of complex networks. Fractal geometry was invented to describe the quantitative aspects of phenomena with multiple scales (no characteristic scale), but what is needed now is a way to describe the dynamics of such complex phenomena/networks. Network characteristics cannot be deduced from the properties of individual components; they must emerge during the formation and growth of the network. Consequently we need mathematics to quantify the strength of the interactions between the network components, as well as to describe how a network develops in time and responds to perturbations (excitations). This has been done through the construction of the decision-making model (DMM) that for very weak coupling is much like the Ising model of cooperative behavior, but for strong coupling can be very different [14], see [8] for an overview.

### 3.1. *Complexity theory and modeling without scales*

One thing we have been studying is how to build a network. The complex adaptive systems and complex adaptive network approaches have been increasingly used in organizational studies. Yet, leadership research has barely begun to use this paradigmatic lens. Given the limitations of the formal theory of traditional leadership and the increasing emphasis on organizational adaptation in asymmetric environments, the complex network approach has potential for providing reasonable gains in leadership research. This research includes the technology for gaining information supremacy while incorporating human-in-the-loop effects to achieve human network integration. Hollingshad *et al.* [15] review of the research done in this area in another of the presentations of this workshop.

Ubiquitous aspects of complex networks are the appearance of non-stationary, non-ergodic, and renewal statistical processes. These properties are manifest through inverse power-law statistical distributions that not only challenge traditional understanding of complexity in physical networks, but require new strategies for understanding how information is exchanged between networks [16], as in the case of interest here between physical and human networks. The approach is to adapt the methods of non-equilibrium statistical physics that have been used to characterize the dynamics of complex phenomena and phase transitions. They were extended to the study of such social phenomena as linguistics, biofeedback techniques and the brain's response to music [17] and to further develop them to model decision-making with incomplete information in an uncertain environment.

A decision maker does not have in mind all possible outcomes, even though that is presented to be the case by positing an 'optimal' strategy. In reality a decision maker is constrained to satisfy a minimum requirement, that being to do the best possible under a given set of constraints or with a given amount of information. This information processing constraint causes problems similar to those arising from time constraints, since an optimal decision is constrained by the time available to make it. Consequently a key issue is to understand the way information is received, assimilated and used, all of which depend on its space-time form.

In order to understand how information is transferred among complex networks such as required in the decision-making process we investigate the phenomenon of memory and examine the influence of memory on stochastic processes. Inverse power-law behavior is known to be characteristic of adaptation, learning, and memory, see for example, Fig. 1. West and Grigolini [18] used renewal theory to develop

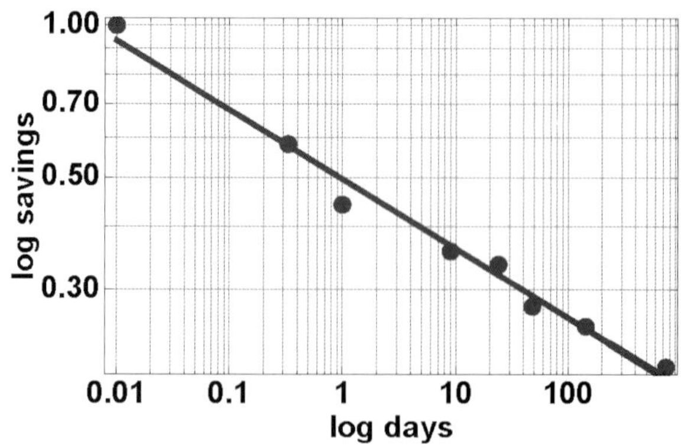


Fig. 1. The closed circles are the data from [19] for the memory saving function and the line segment is the best mean-square fit to that data. The data are clearly well represented by an inverse power-law function (solid line segment) with a slope of –0.14.

a phenomenological model of forgetting that introduced a new psychophysical concept, chipping; discrete events that chip away at chunks of memory and thereby produce forgetting. The neural mechanism producing these chips is the 1/f-noise that is generically produced in complex neuronal networks.

It is the ubiquity of chunking that underlies the universality of the power law form of the memory curve in practice. We have used this observation to introduce the notion of events to account for the loss of memory through the disaggregation of discrete elements from memory chunks. It is important to point out that the abrupt switching off of learning has the effect of triggering the occurrence of a chip, which activates the renewal process, with the resulting chipping away at what was learned. Thus, immediately after we stop learning we start forgetting. This activated forgetting occurs with a forgetting rate given by the power-law index $\mu = 1.86$ so that the mean time for the generation of chips is infinite and the process is not ergodic.

West and Grigolini [20] developed a psychophysical model of decision making based on the difference between objective clock time and the human brain's perception of time. In this model the utility function is given by the survival probability, which is shown to be a generalized hyperbolic distribution. The parameters of the utility function are fit to an inter-temporal choice model whose experimental data and decision making is determined to be a 1/f-noise process, as depicted in Fig. 2.

In the West–Grigolini psychophysical model of decision making the power-law index μ provides a measure of dynamic inconsistency, which is to say how soon individuals change their minds with the passage of time, but without new information. The change in the discount rate over time measures irrationality in terms of the deviation of the generalized hyperbolic (Pareto) distribution from the exponential utility function. The empirical estimates using Takahashi's data [20] show that irrationality in inter-temporal choice is stronger when the outcome is irrelevant to the decision maker, with smaller values of the index $\mu$ corresponding to greater irrationality. It is well known that such generalized hyperbolic delay time distributions have diverging mean times for $\mu < 2$, which appears to be the situation for decision making.

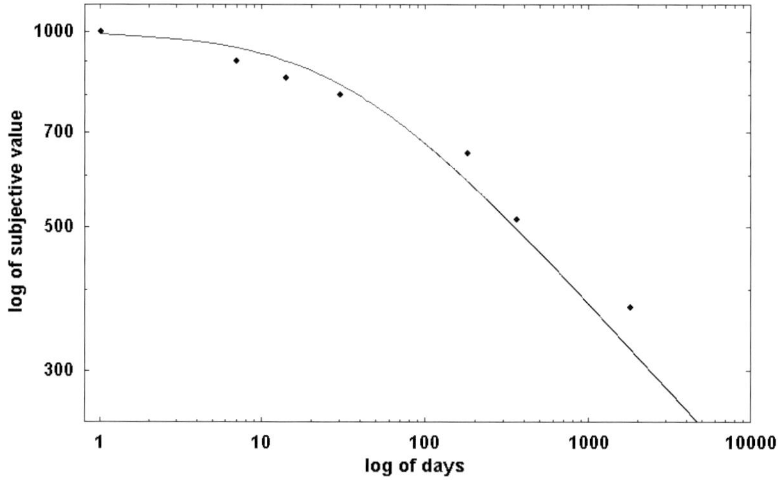


Fig. 2. The logarithm of the utility function is plotted as a function of the logarithm of the delay time. The dots are the experimental points for an individual choosing their own rewards with delay and the solid line segment is the fit of a Pareto distribution to the data. The numerical values of the parameters determined by the least-square fit to the data. [Taken from [20] with permission.]

The exponential form of the discounted utility model rests on a time-independent discount rate. Moreover, even if ordinary people are internally rational, as West and Grigolini assumed, the measured decisions they make in the external world are characterized by the generalized hyperbolic utility function and consequently often appear to the world as irrational.

In a military context this finding, if it is borne out by subsequent experiments, is even more significant. The perceived rationality of others influences such emotional states as trust. A leader may be reluctant to trust his subordinates, and therefore not be willing to delegate authority if the irrationality is suspect. It remains for future studies to examine how the irrationality in the choices people make for others can be mitigated.

### 3.2. *Information propagation in complex adaptive networks*

A perennial problem in the physics of communications is matching information flow to the physical structure of the channel supporting that flow in such a way as to maximize efficiency, that is, maximize the information transferred within the channel. For the purposes of this program it was useful to replace the notion of channel with that of a network in order to avoid unnecessary confusion. To understand the dynamics of complex networks the traditional analysis of information flow and network traffic, involving exponential distributions of messages and consequently Poisson statistics of traffic volume, was abandoned [16].

The investigation into information propagation done here is distinct from what has been done in the past in that we do not use the information entropy of Shannon, nor the control theory of Wiener, but rather we propose to address questions of spatial heterogeneity and temporally non-stationary dynamic complex networks that were almost universally ignored in the past; recognizing that traffic, whether on the Internet or in the brain, is not statistically uniform, but occurs in bursts in both space and time and links are not homogeneous but vary in quality. We investigated the influence of long-time memory, often called non-Markov behavior, or heavy-tailed distributions, on packets in complex communication networks, and their impact on information measures.

The modern mathematical representations of time keeping network nodes are often chaotic attractors that are periodic, but not harmonic, for example, the Rössler oscillator has been used to represent the output of a single neuron [21]. Wood *et al.* [22] have shown that coupled stochastic clocks can manifest cooperative, that is to say, synchronized behavior. This behavior was also found using the DMM where a critical value of the coupling parameter was found to result in a phase transition to a state of consensus [14]. Varela *et al.* [23] postulate that brain function, such as cognition, rests on the cooperative behavior of collections of many neurons. The strategy of Bianco *et al.* [24] is to map brain activity onto networks and consequently study the dynamics of the collections of neurons modeled as such networks. Changes are found in the topology of the network that describes the brain activity as driven by a non-Poisson renewal (NPR) process operating in the non-ergodic regime. Moreover, a collection of blinking quantum dots has the same dynamical behavior as brain activity [25], and so Bianco *et al.* [17] postulate that the two complex networks may share a dynamic model. We further extended this analysis to a cognitive model of decision-making through the exchange of information between complex networks. This was part of the search for common dynamic features among complex networks in various disciplines.

There is general agreement that the brain is a source of 1/f noise [16,26]. This perspective also suggests why 1/f spectra are present in language and in the artwork of the masters, which are a product of the creativity and variability of the brain. Using the PCM [16], understanding why music exerts its spell on the brain [17] becomes evident as well. In fact, it is well known that music generates 1/f spectra and some authors believe in the so-called "Mozart effect," namely that certain types of music may correlate with higher levels of creative and analytic reasoning and more generally with intelligence. The neural foundation of the decision making processes is, at the same time, of fundamental importance for the emerging science of Neuroeconomics.

The brain is probably the most interesting example of a complex network having 1/f variability as determined through the analysis of EEG time series and magnetoencephalogram recordings. The number of mathematical models that have been used to explain the generation of 1/f noise is nearly as broad as the number of disciplines in which the phenomena are observed. A recent strategy for modeling the dynamic variability captured by the 1/f spectrum appears to be independent of the specific mechanisms generating it and involves fractal statistics in one way or another. This approach actually began with fractionally integrated white noise; it was extended to fractal Brownian motion; it was further modified to fractal shot noise; it was shown to be related to the statistical mechanism of subordination; and it was subsequently related to fractal renewal processes. This historical trajectory has now targeted non-the complexity captured in 1/f noise [16]. Grigolini *et al.* [27] developed a theory of 1/f noise of human cognition to explain the recent experimental observations that increasing the difficultly of cognitive tasks accelerates the transition from observed 1/f noise to white noise in decision-making time series.

## 4. Preliminary Conclusions

The conclusions drawn here are preliminary because the research program is only completing its second year. We anticipate that the phenomena we have being able to formally describe for the first time will be understood more completely over the coming year. For example we anticipate being able to establish the conditions under which the principle of complexity matching (PCM) can be proven in a mathematically rigorous way. Moreover we expect this proof to tie together the rigorous mathematics of the generalized fluctuation-dissipation theorem from non-equilibrium statistical physics [8] and the computations of the DMM [14]. From this marriage will emerge a more complete understanding of the mechanisms underlying phase transitions and insight into how consensus is reached in complex social networks.

The research into the decision making theme of the program has been addressed using a variety of strategies. The mathematics of complex networks using the newly developed DMM to understand consensus [14]; the psychophysical model of how individuals make decision and then irrationally change their minds along with the agreement between theory and experiments [20]; how we habituate to the familiar [28] and forget the uninteresting [18]; how the brain processes tasks of increasing complexity during decision making [27]; and finally how all these various pieces fit into the overall picture of exchanging information between complex networks [8,16].

So where does knowledge come into this picture? Knowledge is the interpretation of information that is provided by theory; without theory there is no knowledge. The position of the Earth relative to the Sun at a given time of year is a data point. The sequence of data points made up of daily measurements throughout the year is a data set. The fact that these data fall on an imaginary ellipse in space, was first observed by Kepler, and constitutes information. The information is that the Earth's orbit is an ellipse with the Sun at one of the foci. Knowledge came into being when Newton recognized that the universal law of gravity could explain the information and given one data point Newton's force law could predict other future data points.

Data and information answer the questions: Who? What? Where? and When? But only theory answers the question: Why? Most human thinking is done through pattern identification. A process of comparing information stored in the brain with information being supplied by the environment. Humans make decisions based on the compatibility of these patterns not on our knowledge about them. A baseball batter done not hit a ball because of his knowledge of Newton's Laws; he hits it because the pattern of the incoming ball matches his picture of a home run.

A theory might be based on first principles such the law of universal gravitation. The first principles have to do with the laws of mechanics and the definition of mass. Once these are accepted then the implications are determined by mathematical reasoning. Therefore postulating a given force law for gravity allowed Newton to put in the framework of mechanics and predict the orbit of the Earth around the Sun, the Moon around the Earth, and all the rest of celestial mechanics.

But not all theory has this pristine form. The phenomenological theory of thermodynamics is quite different. Rather than postulating a first principle such as a force law, thermodynamics began with empirical evidence. The first law of thermodynamics is the conservation of energy. The second law of thermodynamics is that isolated physical systems become disordered with the passage of time. The phenomenology of thermodynamics ought to be derivable from first principles and attempts to do this are the subject of statistical mechanics. In statistical mechanics the idealized laws of mechanics are made more 'human' through the introduction of uncertainty, which is to say, through statistical fluctuations. In this way thermodynamic information is transformed into statistical mechanical knowledge in which we know that all data has noise. The uncertainty of noise is understood to come from the environment in simple networks, but can be generated by the nonlinear dynamics in complex networks. Consequently the information in simple networks is found to be constant, but the patterns in complex networks can vary over time.

The major accomplishment of the present research has been the identification of the first universal principle in the science of networks, that being, the principle of complexity management (PCM) discussed in

[8]. PCM states that the maximum information is exchanged between two complex networks when there is compatibility or matching of the complexity of the two networks. A preliminary mathematical proof of this theorem has been made and will be completed in the coming year.

## References


1. *Network Science*, National Research Council of the National Academies, Washington DC, www.nap.edu (2005).
2. *The Implementation of Network-Centric Warfare*, Office of Force Transformation, (2004).
3. J. Garstka and D. Alberts, *Network Centric Operations Conceptual Framework, Version 2.0*, unpublished (2004).
4. B.J. West, *Where Medicine Went Wrong; Rediscovering the Path to Complexity*, World Scientific, Singapore (2006).
5. B.J. West, "Fractal physiology, complexity and the fractional calculus", in *Fractals, Complex Systems*, Eds. W.T. Coffey and Y. Kalmykov, Wiley, New Jersey, 2006.
6. R. Albert, H. Jeong and A. Barabasi, "Attack and error tolerance of complex networks", *Nature* **406**, 378 (2000).
7. J.M. Carlson and J.C. Doyle, "Highly optimized tolerance: a mechanism for power laws in designed systems", *Phys. Rev. E* **60**, 1412 (1999).
8. B.J. West and P. Grigolini, "The principle of complexity management", in this volume.
9. B.J. West, M. Bologna and P. Grigolini, *Physics of Fractal Operators*, World Scientific (2003).
10. P. Essans, M. Spaans and W. Treurniet, "Agile networking in command and control", *The International Journal C2* ***1***, 177–210 (2006).
11. S. Strogatz, *Sync*, Hyperion books, New York (2003).
12. Y. Kuramoto, *Physica A* **126**, 128 (1981).
13. A.T. Winfree, *The Geometry of Biological Time*, Springer-Verlag, Berlin (1990).
14. M. Turalska, M. Lukovic, B.J. West and P. Grigolini, "Complexity and synchronization", *Phys. Rev. E* **80**, 021110 (2009).
15. N.W. Hollingshad, P. Allegrini and P. Grigolini, "Random walk on complex networks: From infinitely slow to the instantaneous transition to equilibrium", in this volume.
16. B.J. West, E. Geneston and P. Grigolini, "Maximizing information exchange between complex networks", *Physics Reports* **468**, 1–99 (2008).

17. S. Bianco, M. Ignaccolo, M.S. Rider, M.J. Ross, P. Winsor and P. Grigolini, "Brain, music and non-Poisson renewal processes", *Phys. Rev. E* (2007).
18. B.J. West and P. Grigolini, "Chipping away at memory", *J. BioCybernetics*.
19. H. Ebbinghaus, Über das Gedächtnis (1885); translated to English as *Memory. A Contribution to Experimental Psychology*, Dover Publ., Mineola, NY (1964).
20. B.J. West and P. Grigolini, "A psychophysical model of decision making", *Physica A* **389**, 3580 (2010).
21. C.J. Stam, *Clinical Neurphyiology* **116**, 2266 (2005).
22. K. Wood, C. Van den Broeck, R. Kawai and K. Lindenberg, *Phys. Rev. Lett.* **96**, 145701 (2006).
23. F. Varela, J.P. Lachauz, E. Rodriguez and J. Martinerie, *Nature Review Neuroscience* **2**, 229 (2001).
24. S. Bianco, E. Geneston, P. Grigolini and M. Ignaccolo, cond-mat/0611035 to appear in *Phys. Rev. E* (2007).
25. M. Nirmal, B.O. Dabbousi, M.G. Bawendi, J.J. Macklin, J.K. Trautman, R.D. Harris, L.E. Brus, *Nature* **383**, 802 (1996); M. Kuno, D.P. Fromim, S.R. Hohmson, A. Gallagher and D.J. Nesbitt, *Phys. Rev. B* **67**, 125304 (2003); K.R. Shimizu, R.G. Neuhauser, C.A. Leatherdale, S.A. Empedocles, W.K. Woo and M.G. Bawendi, *Phys. Rev. B* **63**, 205316 (2001).
26. P. Allegrini, D. Menicucci, R. Bedini, L. Fronzoni, A. Geminani, P. Grigolini, B.J. West and P. Paradisi, "Spontaneous brain activity as a source of ideal of 1/f noise", *Phys. Rev. E*.
27. P. Grigolini, G. Aquino, M. Bologna, M. Luković, B. J. West, "A theory of 1/f noise in human cognition", *Physica A* **388**, 4192 (2009).
28. B.J. West and P. Grigolini, "Habituation and 1/f noise", *Physica A* **389**, 5706 (2010).

# VIEWING THE EXTENDED MIND HYPOTHESIS (CLARK & CHAMBERS) IN TERMS OF COMPLEX SYSTEM DYNAMICS

GERHARD WERNER

*Department of Biomedical Engineering,*
*University of Texas at Austin, TX*
*gwer1@mail.utexas.edu*

In the course of the past 60 years, the brain — and in lockstep with it, Cognition — became liberated from confinement to the skull: the liberation I am referring to consists of the transition from being a distinct physical entity in indirect, mediated contact with the rest of the physical world to being an integral component of it, in a manner envisioned by the *Extended Mind Hypothesis* of Clark & Chalmers: in current terminology, merging brain, body and world into ONE complex system. As background, I briefly review the progression of steps that culminated in the Extended Mind Hypothesis, and allude to the controversies it raised. Assuming the validity of this hypothesis, I will explore the issues that arise from viewing brain, body and world as ONE complex dynamical system. This will lead me to suggest that interrelations between complex system and fractal dynamics enable the seamless integration of human capabilities and the material world.

## 1. Background

Beginning with the 1940s, rapid advances in control, automata, computation and information theories consolidated to an interdisciplinary orientation, in part under the aegis of Cybernetics. This fostered a computational perspective in Neuroscience, extending also into the Social Sciences and Psychology. A signal event was McCulloch & Pitt's proof that binary logical circuits can formally represent the logic of the Principia Mathematics. Impressed by neurons being in a sense binary devices in virtue of their all-or-nothing spike discharges, it was in fact quite tempting to think of the brain as a computing device. The development of artificial neural networks also nourished such a view. In its literal meaning, this view was, however, gradually abandoned by the working investigators, but computational jargon was — and is being-continued in everyday discourse.

In this context, a few words on Epistemic Hygiene may be in order, as a prerequisite for unambiguous discourse across disciplines. Taking it literally, speaking of neural computation in biological systems (or computation in the mind) is of course a gross violation of first principles of logic: Computation is not a Natural kind (in the Philosopher's book). Planet's do not compute their trajectories, they reflect the action of physical forces. I will not belabor this issue since I have written about this on previous occasion (Werner, 2007), and you may say: that's being petty. I submit it is not, and a few words may be appropriate: what had started in the 40s and 50s as a close analogy between electronic computers and brains turned in time into a 'poetic metaphor'. Metaphors can be helpful with finding new perspectives of old problems, or propelling thought into new directions. On the other hand, as "Trojan Horses", they may also canalize thought and communication in some directions and forecloses others. As N. Wiener reportedly once said: "The price of Metaphor is eternal vigilance". What 'computing' in such situations can stand for — and obscures — may in general be a host of physical processes that bring about the particular effect under consideration for which 'computation' captures the outcome, but discounts its Physics. These admonitions apply throughout the ensuing discussion and applies to locutions such as, for instance, 'flow of information', where the metaphorical use can detract attention from the details of physical events in the phase space of systems, perhaps better characterized in neutral terms as perturbations. This issue will be pursued in the Section on 'Is this Plausible?'.

By 1960, the computational framework combined with elements of the Physical Symbol Theory of Newell and Simon to the Computational Theory of Mind (CTM): its central ingredient is the idea that interaction with the world would generate symbolic representation in the brain, on which it would perform problem-solving computations. Its parent discipline is the strictly algebraically rooted Artificial Intelligence. Elegant and in many ways productive (perhaps even seductive) as the framework of the CTM was, its limitations became apparent when adaptive behavior of organisms came under scrutiny. Simon's parable of the ant, told in "The Sciences of the Artificial" (1969), epitomizes the issue: imagine an ant on the beach, as it winds its way through obstacles. Simon contemplated two ways of viewing this situation: one, based on sensory input, the ant creates in its brain a representation of the environment it confronts, and solves the problem by 'computing' an (abstract) plan that would guide the behavior for resolving its difficulties. Note the Cartesian influence of the Mind's indirect, mediated contact with the world. As alternative, the ant would interact with the environment directly — essentially by trial and error and on-line sensori-motor feedback — to maneuver its way through obstacles. Simon chose the former as the paradigm which eventually became the foundation of CTM. The various shapes which the alternative "road not taken" took in the course of the next two decades is the subject of the following introductory comments. After various detours, Simon's ant finally found its way in on-line, causal interactions with the environment, in a dynamical system perspective.

In the evolution to what became called "Second Order Cybernetics", some members of the Cybernetics group turned to problems of self-organization, self-reference, seeking imaginatively, but in vain, to accommodate the observing-acting subject within an expanded cybernetic system. Taking an entirely different approach, Maturana and Varela (1980) applied the theory of Autopoiesis to Cognition, viewing the brain as a closed system, subject to perturbation by external influences and responding to them according to its internal dynamics. An autopoietic system is a member of the class of autonomous self-organizing systems in which the intra-systemic components interact nonlinearly according to purely local rules. In addition, it must also in virtue of its endogenous activity establish and maintain its material unity within the space in which it exists and with which it is structurally coupled.

In the form of the Darwinian Brain, Edelman's (1987) group selection theory transfers some structural aspects of the problematic of adaptive behavior to the brain's ontogeny during which environment appropriate neural structures develop interactively. Extending this principle of structuration by environmental interaction to Cognition became the basis of school of thought of Enactive Cognition (Varela *et al.*, 1991; McGee, 2006 a,b). It postulates the emergence of cognitive structures from recurrent sensorimotor body-environment interactions, which can then subserve perceptual guidance of action. Philosophically, this approach is grounded in the doctrine of Radical Constructivism (Glasersfeld, 1995; McGee, 2005 a,b) and in Autopoiesis (Maturana & Varela, 1980). Employing different robotic platforms, Lungarella and Sporns (2006) demonstrated that sensorimotor interactions and body morphology jointly do indeed induce statistical regularities within the robot control architecture.

In an accommodation to include adaptive behavior into the agenda of the CTM, and under the exigencies of subserving developments in robotics the "Good Old Artificial Intelligence" (GOFAI) augmented its agenda by incorporating features which are entirely alien to the original exclusively representational computational perspective: as Embodied Artificial Intelligence, it came to include object manipulation, locomotion, bodily sensation and feelings (correcting, as it were, "Descartes Error" (Damasio, 1994)) in its agenda: all being 'complex', high-bandwidth, on the fly interactions with the world (Pfeifer & Iida, 2004; Costa-Rocha & Roche, 2005). Two trends towards further extension can be discerned: on the one hand, "Ubiquitous Computing" placing sensors and actuators everywhere, including human bodily organs for regulating their functions, with theory development vastly lagging behind technology; on the other, the subsumption architecture (Brooks, 1991) consisting of a hierarchy of simple reactive, self-organizing modules eliminating the need of any (executive) regulatory coordination of the kind associated with cognition: "the world is the best model of itself".

Capitalizing on the framework of the interdisciplinary field of Synergetics (Haken, 1983), the approach of Coordination Dynamics applied notions of self-organization, metastable brain dynamics and phase transitions initially to regulation of motor behavior (Kelso, 1995). Subsequent extension of these operational properties to dynamical interactions of cortical regions provided a conceptual model for Cognition (Bressler & Kelso, 2001; Kelso, 2003).

A major epistemological shift towards cognition as inseparable from context is associated with the position taken by Situated Cognition (Clancey, 1997; Robbins & Ayede, 2009): although also emphasizing transactional behavior, it views perception and conception as being dynamically formed in concert with the conditions that environment and context provide. It is related to ideas of J.J. Gibson's 'Ecological approach to visual perception" of the 1970s: what is at stake is the organism's capacity for making distinctions in the environment (or prevailing context) such that opportunities for action are identified. The act of grasping an object illustrates the idea: the organism's activity discerns the object's features that afford the opportunity for grasping it. In the conceptual sphere, this principle amounts to acting in real-time and direct (unmediated) interaction with the prevailing material and social circumstances. Intelligent action is on-line and transactional in that it depends in essential ways on the material and social circumstances "on hand', as it were, in direct contact. Drawing on insights from Ethnomethodology, it considers face-to-face communication and conversation as 'ensemble work', and it views behavior of complex organizations as forms of situated communication. In Situated Cognition, the role of the human component of the system is to identify and assess aspects of the (material or social) context for enlisting them for goal achievement: this occurs in direct perceptual-motor contact with the context, with action sequences (rather than abstract plans) being formed 'on-the-fly', opportunistically based on features offered by the environment.

A related emphasis originated with the recognition of the integral co-dependence of perception and action as a complex dynamic system of causal relations. This framework of 'Embodied Cognition' considers cognitive processes rooted in the body's interaction with the world. The realization that 'the brain has a body' (Chiel & Beer, 1997 was primarily based on studying adaptive behavior in natural and simulated conditions in the context of the body's biomechanics. In distinction from Computational Neuroscience and CTM, the field of Computational Neuroethology (Beer, 1990) enlists biomechanics and ecology as essential contributors to enabling adaptive behavior, alongside with neural regulatory control. The extension of this point of view from sensorimotor behavior to sensorimotor accounts of cognition was consolidated to the doctrine of Embodied-Embedded Cognition in Psychology (O'Reagan & Noe, 2001; Noe, 2004; Hurley, 1998). Its claims include several distinct features: cognition is situated, body-based, and action oriented; it favors a 'minimal memory strategy' in virtue of off-loading tasks to the environment (Wilson, 2002). Haugeland (1998) illustrates the essential idea: consider the ability to get to a distant city by car. CTM would postulate a plan or an internal map in your mind. But in this framework, the required information is given in the world in form of the road itself, directing the driver from junction to junction. In this sense, at least some of the conditions for intelligent behavior are "external", provided in the world. This principle is applied in many practical applications, informing human-computer interactions being one of them. The monograph Embodied Grounding (Semin & Smith, 2008) documents social, cognitive and affective aspects of embodied interpersonal relations. This approach influenced the Neurosciences perspective on Cognition (Wheeler, 2005), and — in one version — seeks to apply the methods of Phenomenology to understanding human experience in terms of embodied brain function (Varela *et al.*,1993).

## 2. On the Extended Mind Hypothesis

In radical distinction from the foregoing accounts, The *Extended Mind Hypothesis* of Clark & Chalmers (1998) negates the distinctness of brain, body and world as separate functional entities. In Haugeland's (1998) terms, it stipulates their functional comingling or integralness. Adopting, in addition, a dynamical system perspective, on-line intelligent action is in this framework typically generated through causal interactions in an extended brain-body-environment system. A moment's reflection will remove what may seem strange and esoteric features of this notion: merely consider the use of the many paraphanelia which participate in 'cognitive' activities. Note how we distribute the 'epistemic credit' in certain situations between them and ourselves: think of long multiplications with the use of paper and pencil, slide rule, books and diagrams. Clark & Chalmers propose viewing the human organism in such situations integrated with an external entity, both together constituting a cognitive system as a whole, with all components in the system playing an active causal role. An experience probably familiar to some (many?) of us is the case of "Writing as Thinking": in the perspective of cognitive integration, I refer to the experience that the creation and manipulation of written vehicles is an integral part of our cognitive processing and, in fact, transforms and shapes our cognitive abilities: writing can be viewed as thought in action (Menary, 2007). Along related lines, Cognitive Linguistics alert us to the mapping of bodily experience to abstract (cognitive) domains, for instance in an expression like "send her my warm regards, seamlessly combining a bodily (thermic) sensation with the abstract concept 'regards' and the action of 'sending'. On the same principle, Lakoff & Nunez (2000) suggest that abstract technical entities in Mathematics can viewed as extensions of the structure of human bodily experiences. The slogan then becomes: "thinking ain't all in the head".

Clarifying the '*Extended Mind Hypothesis*' in more detail, Clark & Chalmers (1998; Clark, 2005, 2008) state that this sort of integrated process counts equally as a cognitive, whether or not it is "wholly in the head". Cognitive function is in a sense spread out across the traditional boundaries of brain, body and world. The adherents of this thesis speak of the 'Parity Principle' or the 'continuity thesis' (Menary, 2006), meaning thereby that there is equality between the contribution of what is traditionally called 'internal elements' (grossly meaning 'inside the skin'), and the causal contribution of participating external events. Putting it in different terms, the reasoning goes like this: "if a part of the world functions as a process which, if it were to go on in the head, we would have no hesitation in accepting it as part of the cognitive process, then that part of the world is for that time part of the cognitive process". Accordingly, "for the purpose of the cognitive process, a certain belief (mental state) can be substituted by some physical feature of the environment that is the vehicle for the content of this belief." (Clark, 2008). It is therefore argued that there is no good reason to deny the external contributions the status of cognitive elements.

Innovative as the "*Extended Mind Hypothesis*" (Clark & Chambers, 1998) is, it raised a number of criticisms. A recent workshop organized by Lyre *et al.* (2009) provides an excellent overview of the range of issues raised by the radical break with the traditional neuro-centric view of Cognition this hypothesis entails: a major concern of the majority of discussants is defending the brain's traditional primacy in Cognition, rather than identifying intrinsic flaws or inconsistencies in the hypothesis. Among others, Rupert (2004) argues that intrinsic differences between remembering based on internal memory and on external resources are simply of a different explanatory kind that cannot be bridged by generic account of memory function that the Parity Principle would require. Clark (2008) offers extensive rebuttals in the recent book entitled *Supersizing the Mind.* In rejecting some of the critical comments, Wilson and Clark (2009) explicate at some length the fundamental distinction of Extended from Embedded Cognition: the Parity Principle is intended to emphatically negate any qualitative likeness in the nature of internal and external processes, as is required for Embedded Cognition.

One of the issues for discussion is whether the Extended Mind Thesis would imply an information processing framework. Note that this would assign computational functions to the materiality of those (external) objects that are thought to be part of the cognitive process. But Wheeler (2009) illustrated convincingly that the materiality of the external contributor need not matter, apart from the fact that it must be able to fulfill the required function. Hence, Extended Cognition must embrace multiple realizability, and must be functionalist. A simple illustration makes the point: as far as the cognitive event is concerned, it is inconsequential whether you perform your long-multiplication with paper and pencil, or chalk and blackboard. Accordingly, the Extended Mind Hypothesis is aligned with a particular form of philosophic stance of Functionalism, namely non-computationalist functionalist stance ('Extended Functionalism' in Clark's terminology), as a more general formulation of the canonical thesis, attributed to Putnam, that a systemic state is constituted by the causal relations that it bears to systemic inputs, systemic outputs, and other systemic states, irrespective of their respective material implementation. Wheeler (2009), referring to an example by MacKay *et al.* (1998) offers a telling illustration involving the flight strips(stating airline name, flight number, route, altitude etc) used by Air Traffic Controllers for coordinating their activity: they implement distinct cognitive roles (memory, modeling states of affairs: two planes in conflict for collusion) in a non-computational format, as part of the cognitive architecture of the flight controller.

In the following sections, I will propose an outline of the *Extended Mind Hypothesis* in terms of Complex Systems theory as basis for its role in real-world phenomena. Simon's characterization of certain classes of systems is the point of departure: Simon contrasted systems that are non-decomposable with systems that are (or are nearly) decomposable. In non-decomposable systems, every part communicates with virtually every other part; by contrast, in decomposable systems, most communicative exchanges take place within the subsystems, with interchanges among the subsystems assuming a lesser role. This classification highlights the distinction between situated and embodied-embedded cognition on the one hand, and Extended Cognition, on the other: the former having features of decomposable systems, with the contribution of each subsystem (i.e. the context and the person) amenable to being studied in (relative) isolation; the latter clearly falling into the non-decomposable category in virtue of the 'Parity Principle'.

## 3. Brain and World as ONE Complex Dynamical System

It was not until the mid 90s, perhaps in part energized by the publication of Port & van Gelder's (1995) *Mind in Motion* that eventually a critical mass of investigators began embracing the dynamic paradigm in Neuroscience and Cognition in a systematic manner. Network Science and Complexity, the latter an ill-defined yet useful "fluid" metaphor (Churchland, 2005), were effective catalysts in the Neurosciences as were the impressive data accumulating from Brain Imaging and other neurophysiological investigations. A large array of experimental and theoretical studies has generated a considerable thrust towards characterizing the brain as a high dimensional, complex dynamical system in a state of criticality, exhibiting dynamical phase transitions in response to external stimuli or spontaneous fluctuations, conceptualized along the lines of Complex Dynamical Networks. Additional features are fractal (power-law) scaling and long-range correlations of activity in space and time, the capability of phase-locking of synchronization and of rapid re-configuration in response to external stimuli, analogous to the "avalanches" observed in brain tissue at the mesoscopic level (for a recent review: Werner, 2010). In the following, the abbreviation EMS will designate the complex Brain-World system.

Historically, questions related to complexity were largely under the auspices of Statistical Physics. The scope became enormously enlarged with the rise of Network Theory which, however, in itself, must not be viewed as "a proxy for the theory of Complexity" (Barabasi, 2007). Rather, the behavior of complex systems is as much rooted in network architecture as it is in the dynamical processes taking place on these networks: complex systems behavior must be accounted for by network architecture as well as the dynamics of the processes taking place on it. For consolidating the nomenclature: Complex Systems (synonymous with Complex Dynamical Systems) refer in the following to networks on which dynamical processes in the framework of Statistical Physics take place. Recent publications reflect the current trend in Complex Systems: naming but a few recent signal contributions, consider Holme & Newman's (2006) study of 'Non-equilibrium Phase Transitions in Networks'; the expositions of the 'Statistical Mechanics in Networks' by Park & Newman (2004), 'Critical Phenomena in Complex Systems' by Dorogovtsev *et al.* (2007); and the role of dynamic attractor models for switching patterns of behavior in functional scale-free networks (Bar-Yam & Epstein, 2004). In a similar vein, Boccaletti *et al.* (2006) offer an extensive and detailed overview of 'Structure and Dynamics in Complex Networks'.

At this point, a distinction must be made quite explicit: what is under discussion is that EMS undergoes its evolution as a unit, in accord with its own intrinsic dynamics. This demarcates it sharply from the conditions where two or more separate systems interact in virtue of being coupled, as is the case in the frameworks of situated and embodied cognition. Information exchange between complex networks is governed by the principle of Complexity Matching Effect (for a comprehensive exposition: West *et al.*, 2008).

For the EMS, a different course must be envisioned: the extensive investigations of Bianco *et al.*, 2005; 2007; 2008) identified activity in brain networks as non-ergodic, non-Poisson renewal processes. This precludes transaction among system elements from being "informational" in the sense of being compatible with the premises of Shannon's Information theory. Instead, the Complex System tools of trajectories in phase space and phase transitions, spontaneous symmetry breaking, self-organization and Renormalization Group Transformations (or some form of coarse-graining) are the transformational processes that may but loosely and colloquially be called 'informational'. A distinctive feature of EMS is also the extent of multiscale node heterogeneity (Serrano *et al.*, 2009) of its elements subsumed under ONE system.

The onset of complex behavior is generally associated with establishing long range correlations with fractal scaling (Kim *et al.*, 2007 a,b). Accordingly, EMS would assume in its dynamic evolution the properties of a fractal object with the characteristic attribute of self-similarity (Song *et al.*, 2005; Goh *et al.*, 2006; Serrano *et al.*, 2008). The fractal object of the EMS can then be viewed as a complex dynamic system of heterogeneous composition, with adaptation to the dynamic changes it itself recursively generates. Its fractality underscores the 'memory' function of fractal scaling in the sense of the fractal object's history influencing its ongoing evolution, in an allometric (West, 2009) rather than a homeostatic perspective. As fractal object, it would maintain self-similarity by constantly self-organizing itself to ensure internal self-consistency. A special feature of this operational blueprint is the seamless bridging of two entirely different conceptual frameworks: Complex System on the one hand, and Fractal Systems (in the sense defined by Barnsley, 1999), on the other.

Is this plausible? And what could it imply?

Are there reasons to believe that this hypothetical construct (or aspects related to it) may have any bearing whatsoever on extended human cognition? As a first step, I turn to language processing as one of the most distinctive features of human cognitive processing. In the form of the "*Language of Thought*" (Fodor, 1975), it epitomizes the framework of computation-representation of CTM. However, adopting a dynamical perspective, it became appropriate to view 'representation' in terms of regions of state space, and 'computational rules' as attractors

(Elman, 1995); the dynamic is supplied by recurrent neural networks (RNN). Systems of this kind learn to recognize and generate languages after being trained on suitable examples. Surprisingly, it turned out that the induction of this ability occurs when small network parameter adjustments bring about a phase transition in the neural network's state space. Once in a certain state, machine states for correct recognitions scale with an exponent of 1.4 (Pollack, 1991), based on analysis with the Grassberger-Procacia method (1983). Considering the RNN as a dynamical system, it appears that its trajectories can locate regions in phase space which then support fractal dynamics: putting it in a graphical way, the phase space would seem 'peppered' with regions for fractal dynamics, which can be reached by the trajectories of the complex system's dynamics.

Starting from different premises, Kolen (1994) questioned the often held assumption that RNN learn computational descriptions of formal languages from examples by simulating finite state machines. Instead, he developed the concept that the RNN's state dynamics itself is an Integrated Function System (IFS) (Kolen, 1993). Yet, more is required for functional competence: a support system must be available, in the form of a context interpreting the IFS output in an information-processing way. Here is then an exquisite case of the synergism of fractal and complex system dynamics. Other observations point into similar directions: Levy (2002) obtained evidence that that every point in the hidden layer of the recursive neural network is either itself part of the fractal attractor of the IFS, or has an orbit that "ends on" the attractor in a finite number of steps.

For application to robot programming, Nishimoto and Tani (2004) generated combinatorial action sequences in an approach that shares essential aspects with the principles of the foregoing language-based results. A repertoire of sensory-action sequences for different action plans is represented as trajectories in the state space of the system dynamics; given the starting point of an action plan, a RNN learns which one of the stored action sequences to select for achieving a goal. It turned out that the compositional mapping from initial to goal state is accomplished by the RNN's recursive functions in the form of self-organizing fractal structures.

Essentially, the processes discussed so far associate symbol processing with IFS, taking place in the context of RNN. Jeffrey (1990) achieved the same objective by applying the "Chaos Game" (Barnsley, 1999) to the construction of spatial representations of symbolic sequences: In Jeffrey's approach, this consisted in the transformation of symbolic strings into points in a d-dimensional space. Apart from being a method for representing local and global properties of DNA sequences, this principle is also fruitful in a number revealing ways: under more general conditions, Tino (1999) proved that the scaling properties of the Chaos game establish a direct correspondence between the informational measure of Renyi entropy spectra on the one hand, and the multifractal characteristics of the sequences they represent. In a different approach, Tabor (2000) identified pushdown dynamical automata as a class of computing devices that corresponds to the class of context free grammars whose computation is organized around fractals.

These diverse sources of evidence are indicative of, and support, the general notion of close interrelations between complex dynamics and fractal processes. In answer to the question posed in the heading of this section, I submit that the evidence for close operational relations between nonlinear system dynamics in the symbolic domain and fractals underscores the premise on which the proposed notion of EMS rests. The essential point of this and related observations (e.g.: Stucki & Pollack, 1992; Blair & Pollack, 1997; Levy & Pollack, 2002) is, in the first pace, the graceful integration of complex system phase space with fractal dynamics: this adds plausibility to the notion of EMS sketched in the foregoing. But a still stronger endorsement rests on the fact that dynamic language functions, being quintessential instances of cognitive processes, can be emulated by the synergism of complex system dynamics and fractal dynamics. However, two tantalizing questions remain: one, wherein does the power of a Fractal System lie? Did we merely shift the problem from one to another domain, with the functions of fractality being an act of Faith rather than an established fact? As I concluded in a recent article (Werner, 2010): how fractals and their self-similar structure unpack layers of 'information" for guiding actions across many scales still eludes our comprehension. And second, what exactly is the nature of that synergism between complex system and fractal dynamics?

Clarifying these relations is one of the prerequisites for securing operational plausibility for the Extended Mind Hypothesis.

## 4. Praxis Ahead of Theory

In "Natural Born Cyborgs", Clark (2003) listed examples of remarkable sensory and motor adapatability of organisms. Brain-machine interfaces and motor prostheses are recent illustrative examples. Carmena *et al.* (2003) implanted arrays of electrodes in the monkey's frontal and parietal lobes while learning to move a cursor with a joystick across a computer screen for rewards; after having learned the task, the joystick is disconnected, but the monkey is able to continue the task simply based on the recorded neural activity which is interpreted through an intervening computer for directly controlling the cursor. Another study went one step further and eliminated the need for an external effector altogether (Scherberge, 2009): as a proof of principle, Moritz *et al.* (2008) recorded neural activity of primary motor neurons in macaque monkeys and transformed these signals to signals which directly activated muscles for wrist rotation. I submit that was ordinarily is referred to as Cortical Plasticity can quite naturally be accounted for in the framework of EMS, affording opportunities for generating new states of the integrated brain-body-world system. In the spirit of the *Extended Mind Hypothesis*, the computer that interprets neural activity is part of the cognitive-whole, exploiting both internal and bio-external resources. In accord with the 'Principle of Parity', the use of bio-external resources is thought to be no different from using existing neural or bodily resources. This aspect will gain increasing importance as new "wearable" computing tools and new progress with Cognitive Neural Prosthetics (Andersen *et al.*, 2010) will become available for exploring the limits of Cognitive Extensibility.

## 5. Conclusion

Taking the *Extended Mind Hypothesis* of Clark & Chalmers at its face value, I examine what issues arise if one were to view it in terms of Complex System dynamics. Two principal issues stand out: one, the role of heterogeneity of network nodes directs attention to aspects of system modeling not hitherto adequately clarified; and, second, taking cues from the cognitive models of computations language processing, what is the significance of the relation between network dynamics and the dynamics of Integrated Function Systems? I submit that a deeper understanding of the functional interdependence and synergy of complex systems and fractality, suggested by their role for symbolic processing in language, will provide the foundation for advancing and enriching the Praxis of Extended-Mind-inspired applications.

## References


1. R.A. Andersen, E.J. Hwang, G.H. Mulliken, *Annu. Rev. Psychol.* **61** (2010).
2. A-L. Barabasi, *IEEE Control System Magazine* **27**(4), 33 (2007).
3. M.F. Barnsley, *Fractals Everywhere*, Academic Press NY, 2nd edition (1993).
4. Y. Bar-Yam, I.R. Epstein, *Proc. Nat. Acad. Sci. USA* **101**(13), 4341 (2004).
5. R.D. Beer, *Intelligence as Adaptive Behavior: An Experiment in Computational Neuroethology*, Academic Press NY (1990).
6. S. Bianco, E. Geneston, P. Grigolini, M. Ignaccolo, *Physica A* **387**, 1387 (2008).
7. S. Bianco, P. Grigolini, P. Paradisi, *J. Chem. Physics* **123**, 174704 (2005).
8. S. Bianco, M. Ignaccolo, M.S. Rider, M.J Ross, P. Winsor, P. Grigolini, *Physical Rev. E* **75**, 061911 (2007).
9. A.D. Blair, J.B. Pollack, *Neural Computation* **9**, 1127 (1997).
10. S. Boccaletti, V. Latora, Y. Moreno, M. Chavez, D-U. Hwang, *Physics Reports* **424**, 175 (2006).
11. S.L. Bressler, J.A.S. Kelso, *Trends in Cogn. Sci.* **5**, 26 (2001).
12. R. Brooks, *Artificial Intelligence* **47**, 139 (1991).
13. J.M. Carmena, M.A. Lebedev, R.E. Crist, J.E. O'Doherty, D.M. Santucci, D.F. Dimitrov, P.G. Patil, C.S. Heniquez, M.A.L. Nicolelis, *PLoS Biology* **1**(2), 193 (2003).
14. H.J. Chiel, R.D. Beer, *Trends in Neuroscience* **20**(12), 553 (1997).
15. P.M. Churchland, *Inquiry* **48**(5), 464 (2005).
16. W.J. Clancey, *Situated Cognition*, Cambridge University Press (1997).
17. A. Clark, D. Chalmers, *Analysis* **58**(1), 7 (1998).
18. A. Clark, *Natural-born Cyborgs: Minds, Technologies and the Future of Human Intelligence*, Oxford University Press (2003).
19. A. Clark, *Analysis* **65**, 1 (2005).

20. A. Clark, *Supersizing the Mind*, Oxford University Press (2008).
21. F.A. Costa, L.M. Rocha, *Artificial Life* **11**, 5 (2005).
22. A.R. Damasio, *Descartes' Error*, GP Putnam (1994).
23. S.N. Dorogovtsev, J.F. Goltsev, F. Mendes, *Rev. Mod. Physics* **80**, 1257 (2007).
24. G. Edelman, *Neural Darwinism: The Theory of Neural Group Selection*, MIT Press (1987).
25. J.L. Elman, *Language as a Dynamical System*, in: Port, R.F., van Gelder, T., Mind as Motion, MIT Press (1995).
26. J. Fodor, *Language of Thought*, Harvard University Press (1975).
27. E. Glasersfeld von, *Radical Constructivism: A Way of Knowing and Learning*, Falmer Press, London (1995).
28. K-I. Goh, G. Dalvi, B. Kahng, D. Kim, *arXiv:cond-mat/0508332v2 [cond-mat-stat-mech]* (2006).
29. P. Grassberger, I. Procacia, *Physica D* **9**, 189 (1983).
30. H. Haken, *Synergetics*, Springer, Berlin (1982).
31. J. Haugeland, J. *Mind Embodied and Embedded*. In: Having Thought: Essays in the Metaphysics of Mind, Harvard University Press (1998).
32. P. Holme, M.E.J. Newman, *Phys. Rev. E* **74**, 056108 (2006).
33. S.L. Hurley, *Consciousness in Action*, Harvard University Press (1998).
34. H.J. Jeffrey, *Nucleic Acid Res.* **18**(8), 2163 (1990).
35. J.A.S. Kelso, *Dynamic Patterns*, MIT Press (1995).
36. J.A.S. Kelso, *Cognitive Coordination Dynamics*, in: The Dynamical Systems Approach to Cognition, eds. W. Tschacher & J-P. Dauwalder, World Scientific (2003).
37. J.S. Kim, K.I. Goh, B. Kahng, D. Kim, *Chaos* **17**, 026116 (2007a).
38. K.S. Kim, K.I. Goh, B. Kahng, D. Kim, *New J. Physics* **9**, 177 (2007b).
39. J.F. Kolen, *Recurrent Networks: State Machines or Iterated Function Systems*, Proceedings of the 1993 Connectionist Models Summer School (1993).
40. J.F. Kolen, *Advances in Neural Information Processing Systems* **6**, 501 (1994).
41. G. Lakoff, R. Nunez, *Where Mathematics Comes From*, Basic Books, NY (2000).
42. S.D. Levy, *Infinite RAAM: Initial Investigations into a Fractal Basis for Cognition*, Ph.D. Thesis, Brandeis University, July 2002.
43. S.D. Levy, J.B. Pollack, *Logical computation on a fractal neural substrate*, http://www.demo.cs.brandeis.edu/papers/raam-ijcnn01.ps.gz (2002).

44. H. Lyre, I. Wachsmuth, S. Walter, http://www.uni-bielefeld.de/philosophie/extendedmind/ (2009).
45. M. Lungarella, O. Sporns, *PLoS Computational Biology* **2**(10), e144 (2006).
46. W.E. Mackay, A.L. Fayard, L. Frobert, L. Medini, *Conf. Proceedings on Human Factors in Computing Systems (CHI 1998)*, New York, ACM Press, pp. 558.
47. H.R. Maturana, F.J. Varela, *Autopoiesis and Cognition: the Realization of the Living*, Dordrecht: Boston Studies in the Philosophy of Science, Vol. 42, D. Reidel (1980).
48. W.S. McCulloch, W. Pitts, *Bulletin of Math. Biophysics* **5**, 115 (1943).
49. K. McGee, *Constructivist Foundations* **1**, 19 (2005a).
50. K. McGee, *Constructivist Foundations* **1**, 73 (2005b).
51. R. Menary, *Philosophical Psychology* **19**(3), 329 (2006).
52. R. Menary, *Language Sciences* **29**, 621 (2007).
53. C.T. Moritz, C.I. Perlmutter, E.E. Fetz, *Nature* **456**, 639 (2008).
54. R. Nishimoto, J. Tani, *Neural Networks* **17**, 925 (2004).
55. A. Noe, *Action in Perception*, MIT Press (2004).
56. J.K. O'Reagan, A. Noe, *Behavioral & Brain Sciences* **24**(5), (2001).
57. J. Park, M.E.J. Newman, *Phys. Rev. E* **70**, 066117 (2004).
58. R. Pfeifer, F. Iida, In: Embodied Artificial Intelligence, *LNAI 3139*, pp. 1–26, Springer Berlin (2004).
59. J.B. Pollack, *Machine Learning* **7**, 227 (1991).
60. R.F. Port, T. van Gelder, *Mind as Motion*, MIT Press (1995).
61. P. Robbins, M. Ayede, *The Cambridge Handbook of Situated Cognition*, Cambridge University Press (2009).
62. R. Rupert, *J. Philosophy* **101**(8), 389 (2004).
63. H. Scherberger, *Curr. Opinion in Neurobiol.* **19**, 1 (2009).
64. G.R. Semin, E.R. Smith (eds.), *Embodied Grounding: Social, Cognitive, Affective, and Neuroscientific Approaches*, Cambridge University Press (2008).
65. M.A. Serrano, D. Krioukov, M. Boguna, *arXiv:o710.2092v2[cond-mat.dis-nn]* (2008).
66. M.A. Serrano, M. Boguna, A. Vespigagni, *Proc. Nat. Acad. Sci. USA* **106**(16), 6483 (2009).
67. H. Simon, *The Sciences of the Artificial*, MIT Press (1996).
68. C. Song, S. Havlin, H.A. Makse, *Nature* **433**, 392 (2005).

69. D.J. Stucki, J.B. Pollack, *Fractal (reconstructive analogue) memory.* Proceedings of the 14th Annual Conference of the Cognitive Science Society, Bloomington, IN (1992).
70. W. Tabor, *Expert Systems* **17**, 41 (2000).
71. P. Tino, *IEEE Transaction on Systems, Man and Cybernetics* **29**(4), 386 (1999).
72. P. Tino, M. Koteles, *IEEE Transaction on Neural Networks* **10**(2), 284 (1999).
73. F.J. Varela, E. Thompson, E. Rosch, *The Embodied Mind: Cognitive Science and Human Experience*, MIT Press (1993).
74. G. Werner, *BioSystems* **87**, 82 (2007).
75. G. Werner (2010). *Frontiers in Physiology*, doi:10.3389/fphys.2010.00015.
76. B.J. West, *Control from an Allometric Perspective.* In: Progress in Motor Control, ed. D. Sternad, Springer (2009).
77. B.J. West, E.L. Geneston, P. Grigolini, *Physics Reports* **468**(1–3), 1 (2008).
78. M. Wheeler, *Reconstructing the Cognitive World*, MIT Press (2005).
79. M. Wheeler, *Minds, Things and Materiality.* In: The Cognitive Life of Things: Recasting the Boundaries of Mind, (eds.) C. Renfrew & L. Malafouris, Cambridge (2009).
80. M. Wilson, *Psychonomic Bull. & Rev.* **9**(4), 625 (2002).
81. R.A. Wilson, A. Clark, *How to Situate Cognition: Letting Nature Take Its Course.* In: P. Robbins, M. Ayede, The Cambridge Handbook of Situated Cognition, Cambridge University Press (2009).

# UNCERTAINTY IN PSYCHOPHYSICS: DERIVING A NETWORK OF PSYCHOPHYSICAL EQUATIONS[1]

KENNETH H. NORWICH

*Institute of Biomaterials & Biomedical Engineering, University of Toronto, 164 College Street, Toronto, Ontario, Canada, M5S 3G9*

To perceive, and more particularly to sense, is conditional upon the state of uncertainty of the perceiving system. This idea is developed from its philosophical base, to the equations that encode it, and through to the experimental data that provide its support. Much of the paper constitutes a review of research conducted over past years, but viewed now with some perspective. This uncertainty concept represents a kind of thermodynamic principle, which tells us how various sensory phenomena will occur, but does not provide a physiological mechanism. Hence, for example, the mathematical structure of the Weber-Fechner law will emerge, but it remains for mechanisms such as, perhaps, neural networks, to define exactly how the law will be implemented physiologically. The latter part of the paper deals with the attempt to link sensation, for one special case, to the physical entropy of the received signal.

## 1. Introduction

The year 2010 marks one-third of a century since publication of my first paper on this subject. It has been a mathematical journey through the many dimensions of sensory science. Strangely, this is the first review of this network of papers that I have ever attempted. This paper is brief, and will emphasize my philosophy of the process of sensation, the gist of the mathematical structure I have tried to create, and the scope of this mathematical structure. I give no complete derivations, but rather just an overview, indicating what sort of equations I derived, and what their significance might be.

---

[1]Much of this work has been supported by operating grants from the Natural Sciences and Engineering Research Council of Canada.

The first component of my work, extending from 1977 to 1992, essentially demonstrated that all of the core empirical or phenomenological equations of psychophysics, from 1865 to the present era, can be derived from two seminal or root equations. This material was summarized in a book, *Information, Sensation and Perception*, published in 1993 [1], and is now available on the internet. The second component of my work, extends to the present time, and is largely concerned with binding the empirical laws of psychophysics, laws such as those of Weber and Fechner, to the laws of physics. How well I have succeeded in these endeavors is, of course, for you to decide.

## 2. Philosophical Underpinnings

My philosopher of reference is George Berkeley, who flourished in the early 18th century, principally at Trinity College in Dublin. In his opus, *A Treatise Concerning the Principles of Human Knowledge*, Berkeley expounded his famous dictum, *esse* is *percipi*: to be is to be perceived. Although seldom credited with being so, Berkeley was an early exponent of the principle of relativity. In his famous treatise, Berkeley wrote quite definitively, "…it doth not appear to me, that there can be any motion other than *relative*; so that to conceive motion, there must be at least conceived two bodies, whereof the distance or position in regard to each other is varied…" I believe that the principle of relativity pervaded the mind of Berkeley, and that his principle of perception was none the less relative in nature. Existence itself was relative to the mind of the perceiver (a position that Einstein never endorsed). I labored for some years in the attempt to incorporate at least a Berkeleian posture into psychophysics or into the physiology of the senses, settling finally on the principle that *to perceive is to doubt or be uncertain*. That is, to perceive, we must be uncertain, and with certainty, perception vanishes. Perception thus becomes relative to the individual who perceives.[2]

---

[2]If existence is contingent upon perception (esse est percipi), and perception contingent upon uncertainty, then uncertainty is necessary for existence of an object — a position with which Heisenberg may not have been unhappy.

## 3. Mathematical Representation of the Psychophysical Law (Weber-Fechner and Stevens)

The question was how to incorporate the concept of uncertainty into the process of sensing. I longed to use Boltzmann's relationship, $S = k \log W$, where $W$ in the number of microstates, but this equation is not easily transportable into physiology or psychophysics. Unbeknown to me in those years was that Abraham Moles (1966) [2] had already observed the similarity between Boltzmann's equation and Fechner's law, which he wrote as $\gamma = \kappa \log(\beta/b)$. But no matter ... I could not obtain a working perceptual model using Boltzmann directly. So I defaulted to a position that looked superficially like that of Shannon's information theory.

Steady sensory signals, which are the only stimuli I shall deal with here, are always accompanied by fluctuations at the microscopic level. For example, tasted solutions consist of fluctuating particle densities at the level of taste receptors; photons (of low intensity beams) impinging on the photoreceptors of the retina are governed by Poisson statistics, etc. So uncertainties are present at the microscopic level. It was argued that uncertainty about the mean value of the sensory signal may be represented by the difference between two Shannon-type entropies in the form $\frac{1}{2}\ln 2\pi e(\sigma_S^2 + \sigma_R^2)$ (due to signal plus reference variances) and $\frac{1}{2}\ln(2\pi e \sigma_R^2)$ (residual reference variance). To obtain the variance of the mean signal, variance was divided by $m$, the number of samples taken by a receptor of its input signal.

$$H = \frac{1}{2}\ln(1 + \sigma_S^2/m\sigma_R^2)$$

For initial study, $m$ was taken to be proportional to the time, $t$, that the sensory signal was in place, and $\sigma_R^2$ was taken as constant, leaving an expression for uncertainty, $H$, in the form

$$H = \frac{1}{2}\ln(1 + b\sigma_S^2/t)$$

where $b$ is a proportionality constant. The incorporation of a time function into the $H$-function was derived differently recently by Medina

[3]. The above expression for $H$ cannot really be tested experimentally, because signal variance is usually not known. It was necessary to append an additional (and final) assumption to the effect that signal variance varies as signal mean raised to a power: $\sigma_S^2 \sim I^n$, where $I$ is the signal mean, and $n$ is a constant. The final form of the $H$-function is, then,

$$H = \frac{1}{2} \ln (1 + \beta I^n / t) \tag{1}$$

where $\beta$ is a constant (distinct from Fechner's $\beta$, shown above). Introducing, now, the philosophical premise that to perceive is to be uncertain, one obtains

$$F = kH \tag{2}$$

where $F$ is a psychophysical variable such as magnitude of sensation, and $k$ is a scaling constant.

Equation (1) is an expression of Shannon entropy, but modified to fit the circumstances. $H$ will be maximal for some minimum value of $t$, corresponding, say, to one or two samples of the input signal. It will then decline with increasing time, representing the process of sensory adaptation.

Adaptation is viewed, accordingly, as a decrease in uncertainty as a signal is sampled progressively with time.[3]

Equation (1) is an equation giving the Shannon entropy at a sensory receptor as a function of the time of exposure to a steady sensory signal. As such, the decrease in $H$ with time gives the information received. We know from extensive experiments performed in the 1950's and 1960's the maximum information, $\Delta H_{\max}$, which can be received from such a steady signal, and this value will be introduced below.

Equations (1) and (2) together with the concept of $\Delta H$ as a quantity of information are, essentially, all that is needed to derive the host of phenomenological equations produced by psychophysics from 1865 to the present. For sensory signals of constant duration, $t_c$, setting $\beta / t_c = \gamma$, we have from Eqs. (1) and (2)

---

[3]Boltzmann's entropy, of course, is nondecreasing with time, under action of random processes. Perceptual entropy is nonincreasing with time, as certainty is gained progressively.

$$F = \frac{1}{2} k \ln(1 + \gamma I^n) \tag{3}$$

For larger mean signal values, $\gamma I^n \gg 1$ and we retrieve Fechner's law, $F = \frac{1}{2} kn \ln I + \text{const}$. For smaller mean signals, $\gamma I^n \ll 1$, and from a Taylor's series we obtain Stevens' law, $F = \frac{1}{2} k\gamma I^n$. These are the two famous forms of the "psychophysical law". The exponent, $n$, is the well-known power function exponent, which takes on specific values for different modalities of sensation. For example, for hearing at 1000 Hz, $n = 0.6$ when $I$ is measured in rms sound pressure units.

## 4. A Network of Equations Issuing from the Entropic Form of the Psychophysical Law

### 4.1. *The differential threshold (ΔH from Fechner's conjecture) and Weber's fraction*

Fechner's concept of the just noticeable difference is illustrated schematically in Figure 1. Stimulus intensity, $I$, can increase continuously with no change in subjective apprehension of the stimulus until intensity has reached a threshold value of $I + \Delta I$. When the threshold is traversed, subjective apprehension increases from its basal value of $F$ to $F + \Delta F$.

The *Weber fraction* is the ratio of a just noticeable difference in signal intensity, $I$, to the "pedestal intensity": $\Delta I/I$. The quantity $\Delta I$ is sometimes called the *differential threshold*. Because $\Delta I \ll I$ usually, we can estimate its value by differentiating Eq. (3) and replacing d$I$ by $\Delta I$. Replacing $F/k$ by $H$, differentiating and rearranging,

$$\Delta I/I - (2\Delta H/n)[1 + 1/(\gamma I^n)] \tag{4}$$

Characteristically, the Weber fraction is large for small values of $I$, and may approach a constant value, $2\Delta H/n$, for larger values of $I$.

Employing Fechner's hypothesis that $\Delta F = k\Delta H$ is constant for one just noticeable difference, Eq. (4) gives the characteristic shape of the curve obtained when Weber fraction, $\Delta I/I$, is plotted against $I$ (Fig. 2). The shape of the function varies somewhat depending on how the Weber fraction is measured (see Wong and Figueiredo [4]), but

Eq. (4) with constant $\Delta H$ is a useful guide, and has been observed over the years for many sensory modalities (e.g. Riesz at Bell Labs for audition [5], Lemberger [6] for sweetness of sucrose). Remember that Eq. (4) may be less accurate for larger values of $\Delta I$.

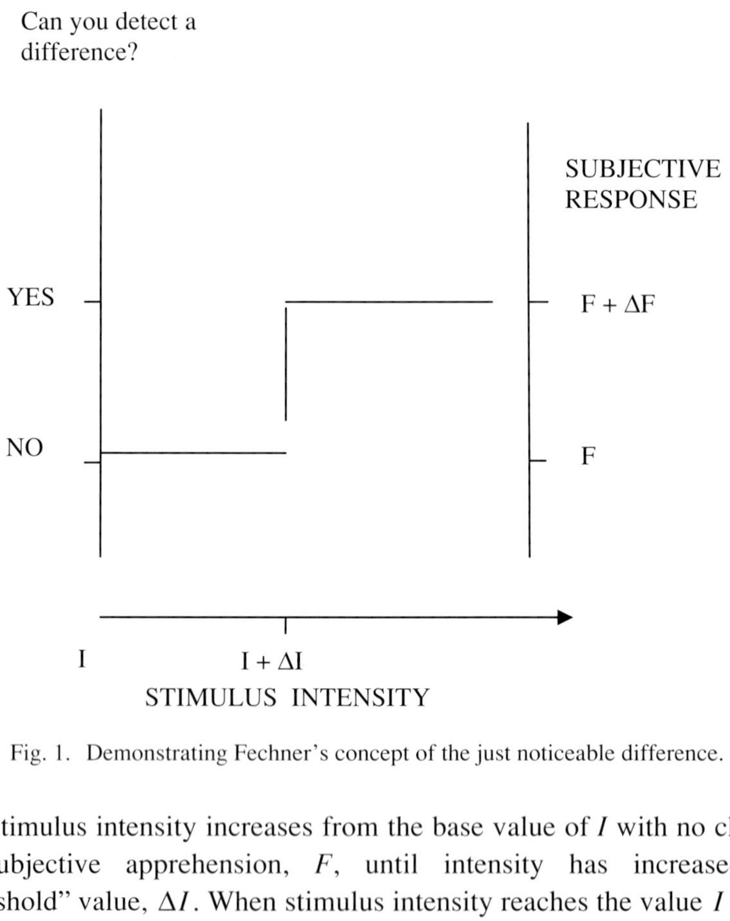


Fig. 1. Demonstrating Fechner's concept of the just noticeable difference.

Stimulus intensity increases from the base value of $I$ with no change in subjective apprehension, $F$, until intensity has increased by "threshold" value, $\Delta I$. When stimulus intensity reaches the value $I+\Delta I$, subjective apprehension jumps by $\Delta F$ to $F+\Delta F$. Although $\Delta I$ will increase as $I$ increases (in accordance with Weber's law), $\Delta F$ will not change.

Weber fraction, $\Delta I/I$, plotted against stimulus intensity, $I$. Eq. (4) has been fitted to the data of Lemberger. Parameter values are as follows: $2\Delta H/n = 0.140$, $\gamma = 1.41$, $n = 1.23$.

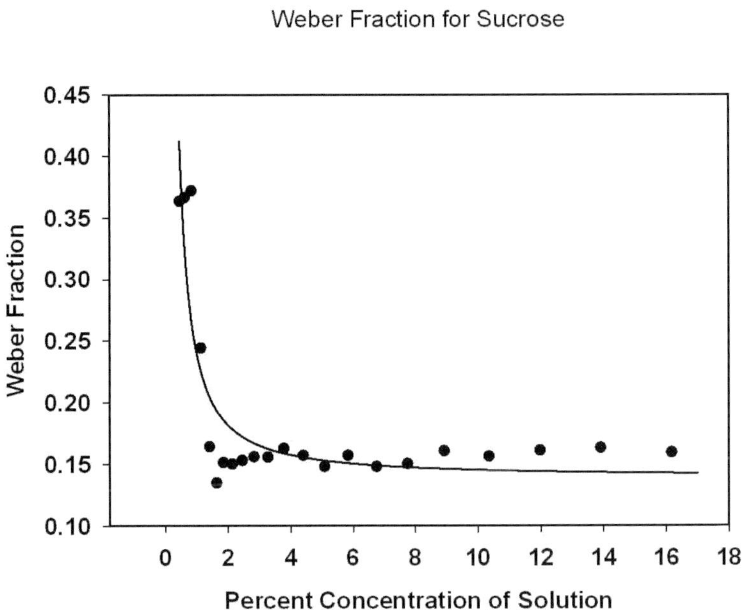


Fig. 2. The shape often, but not invariably, assumed by the Weber fraction.

A more modern treatment of the Weber fraction is given by Wojtczak and Viemeister [7]. The adaptation of the entropy equation (4) to capture the various experimental protocols described by these authors is given by Wong and Figueiredo [4].

### 4.2. *The hyperbolic law governing the magnitude of n ($\Delta H$ from Miller's magical number)*

Employing Eqs. (2) and (3) in the Fechnerian format, we have

$$H = \frac{1}{2} n \ln I + \text{const.} \tag{5}$$

Let $I_1$ approximately equal the highest intensity value within the physiological range for some modality of sensation (e.g. about 1 Molar for a tasted solution; about $10^5$ N/m$^2$ SPL rms for sound, etc). Let $I_2$ approximately equal the lowest intensity value for which the Fechner approximation of the entropic equation is valid. Substituting these values into Eq. (4) and subtracting, we obtain

$$\Delta H \equiv H_2 - H_1 = \frac{1}{2} n \ln (I_2/I_1)$$

from which

$$n \cdot \log_{10} \Re = 2\Delta H/\ln 10 \tag{6}$$

where $\Re$, the range, is defined to be $\log_{10}(I_2/I_1)$. It is known from experiment that the value of $\Delta H$, the maximum drop in stimulus uncertainty (or maximum information obtainable from a steady stimulus) is approximately equal to the natural logarithm of Miller's "magical number" 7. One never quite obtains a value as great as 7, and 6 is a better estimate of this quantity,[4] so that Eq. (5) yields

$$\mathrm{n} \cdot \log_{10} \Re \cong 1.56 \tag{7}$$

That is, the variables $n$ and $\log_{10} \Re$ tend to form a rectangular hyperbola whose parameter equals 1.56. This result has been confirmed experimentally.

One final example using the concept that $\Delta H$, the increment in information, provides the limiting quantity needed for perception is the simple reaction time.

## 4.3. *Simple reaction time (ΔH is the minimum quantity of information needed to react)*

Simple reaction time, $t_r$, is the time interval from onset of a sensory signal (e.g. light, sound) to the earliest response to the signal (e.g. pressing a button). Define $t_o$ as the value of $t$ for which $H$ is maximum (corresponding, perhaps, to $m = 1$). Keeping $I$ constant, we can evaluate a critical minimum value $\Delta H$ equal to the information required to react. From Eq. (1),

$$\Delta H = \frac{1}{2} \ln (1 + \beta I^n/t_o) - \frac{1}{2} \ln (1 + \beta I^n/t_r) \tag{8}$$

This equation can be solved for $t_r$ giving the cumbersome expression,

$$t_r = [1/(t_o e^{2\Delta H}) - (1 - e^{-2\Delta H})/\beta I^n)]^{-1}$$

[4]One might note for later reference that ln 6.36 = 1.85.

However, it is readily recognized that the minimum possible reaction time, $t_{\mathrm{r\,min}}$, is given by

$$t_{\mathrm{r\,min}} = t_{\mathrm{o}} \mathrm{e}^{2\Delta H} .$$

Similarly, one can solve for $I_{\min}$, the minimum value of signal intensity, $I$, for which a response is possible. After a modicum of algebra, one can show that simple reaction time can be expressed by

$$t_{\mathrm{r}} = t_{\mathrm{r\,min}} / [1 - (I_{\min} / I)^n] \tag{9}$$

The two equations derived for $t_{\mathrm{r}}$ are mathematically identical, however, from Eq. (9) one sees that only three parameters, $t_{\mathrm{r\,min}}$, $I_{\min}$ and $n$ are needed to specify simple reaction time. From Eq. (9) we also see that larger signal values, $I$, will give rise to shorter reaction times, a feature of simple reaction times that had been observed experimentally since the late nineteenth century. This equation fitted to vision data is shown in Fig. 3. It is also rewarding to approximate the expression for $t_{\mathrm{r}}$ by expanding the denominator in a binomial series:

$$[1 - (I_{\min} / I)^n]^{-1} = 1 + (I_{\min} / I)^n + \cdots$$

retaining only the first-order term. Hence Eq. (9) becomes

$$t_{\mathrm{r}} = (t_{\mathrm{r\,min}} I_{\min}{}^n) I^{-n} + t_{\mathrm{r\,min}} , \tag{10}$$

which is the form of the equation for simple reaction time suggested by Piéron in the early 20th century, but which may also be found in his textbook [8].

The dates for the discovery of these empirical relations are old. We are demonstrating here that most of these vintage phenomenological rules are really just special cases on the general law of entropy expressed by Eqs. (1) and (2), and constrained as necessary to match the experimental conditions governing the phenomenological rules. There are many other phenomenological laws and empirical observations that can be derived. They are summarized in flow charts (Norwich, [1] pp. 248–249).

$$t_{\mathrm{r\,min}} = 0.117 \text{ s}, I_{\min} = 0.449 \text{ relative units}, n = 0.439 .$$

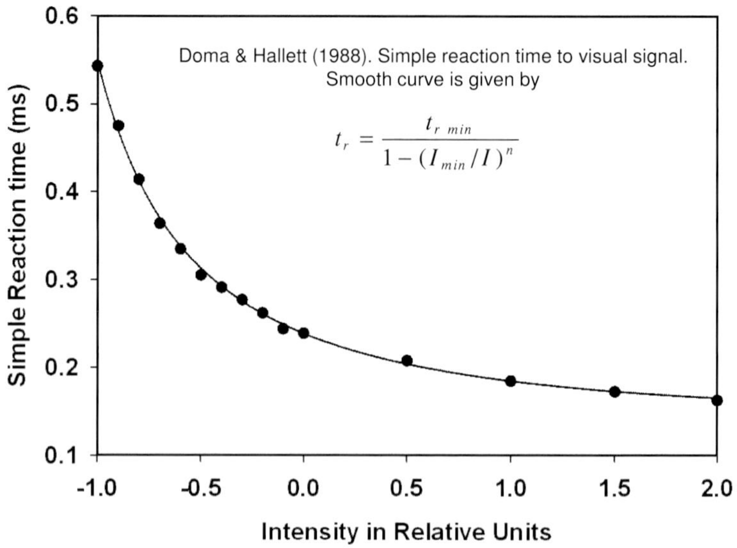


Fig. 3. Data of Doma and Hallett [9] showing the decline in simple reaction time to a visual stimulus with increase in stimulus intensity. Equation (9) has been fitted to the data.

## 5. Searching for Support within Thermodynamics and Statistical Physics

Perhaps one should be content with Eqs. (1) and (2) which do seem to have some scope. If there is a problem with these equations it is that they seem to have been created *ex nihilo*, presenting as a new principle in sensory science (sensory magnitudes as measures of uncertainties, Eq. (2)) and a new way of expressing these uncertainties (Eq. (1)). I have tried to replace the second of these novelties, Eq. (2), by appeal to classical physics.

Equation (2), $F = kH$, is parallel to the statistical mechanical relation, $S = k_B H_E$, where $S$ is the thermodynamic entropy, $k_B$ is Boltzmann's constant and $H_E$ is an uncertainty function governing molecular energy (e.g. [10]). That is, $H_E$ is of the form

$$H_E = -\Sigma\, \mathrm{p}(E_\mathrm{i}) \ln \mathrm{p}(E_\mathrm{i})$$

where $\mathrm{p}(E_\mathrm{i})$ is the probability of energy $E_\mathrm{i}$. To this end, I made a preliminary exploration of the concept that psychophysical magnitude is a direct expression of the *physical* entropy of the signal being perceived. As with all theoretical equations discussed here and elsewhere, I proceeded only if there was experimental evidence to support or negate the concept. The first sensory system that I discovered to have adequate experimental support is the sensation of the taste of saltiness. The supporting data emerged from German scientific literature of the 1920's.

## 5.1. *Emergence of the Weber–Fechner law from thermodynamics*

The following is a brief outline of the physical chemistry needed to explore the psychophysics of the taste of saltiness of chloride salts, which are strong salts. I shall try to give just enough detail to convey the flavor of the argument, leaving it to the reader to explore further [11]. An intriguing corollary of this work is the appearance of the gas constant, *R*, within the mathematical expression of the psychophysical law.

A strong (1:1) electrolyte, such as a chloride salt, will dissociate in solution into two monovalent ions:

$$\mathrm{AB} \rightarrow \mathrm{A}^{+} + \mathrm{B}^{-}$$

Then the partial molar entropies of the cation and anion are given, respectively, by

$$S_{+} = S_{+}^{o} - R \ln a_{+}$$
$$S_{-} = S_{-}^{o} - R \ln a_{-}$$

where $S_{+}^{o}$ and $S_{-}^{o}$ are hypothetical standard molar entropies, $a_{+}$ and $a_{-}$ are activities, and $R$ is the gas constant. Adding these equations, and setting

$$S = S_{+} + S_{-},\ \ S^{o} = S_{+}^{o} + S_{-}^{o},\ \text{ and }\ a = \sqrt{(a_{+}a_{-})}\,,$$

we obtain

$$-S = -S^{o} + 2R \ln a\,, \tag{11}$$

relating molar entropy, $S$, to the logarithm of activity, which in turn, reflects the concentration of a solution to be tasted.

In order to represent Fechner's law within the physical chemistry, we must allow for the "absolute threshold" of the taste of saltiness, the

no other instance where the value of a physical constant can be estimated directly from psychophysical data.

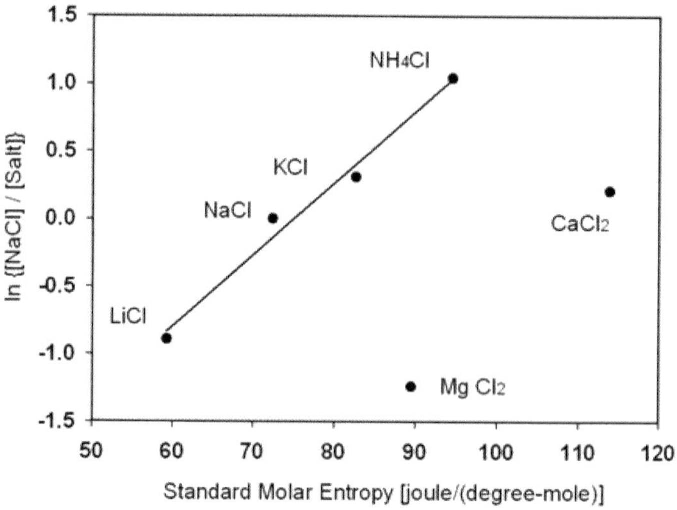


Fig. 4. Data of von Skramlik [12]. Logarithm of the ratio of the concentration of sodium chloride to the concentration of various chloride salts that tasted equally salty, plotted against the standard molar entropy of the comparison salt. The data separates into two sets. The (1:1) salt data is arrayed linearly, and the regression line through these data points is shown. The (2:1) salt data (two points only) lie below this line.

## 6. Discussion

### 6.1. *Review*

I have reviewed somewhat cursorily rather a broad area of my sensory research — the first such review I have written — extending from my first published work on the subject in 1977 up to 2001. The first part dealt with the philosophical base for the mathematical studies that followed: the principle that uncertainty is necessary for sensation (and perception) to occur. This principle was encoded in the equation, $F = kH$, where $F$ is the perceptual variable and $H$ an uncertainty function. $H$ was then modeled for simple sensation in a manner such that the magnitude of the function increased logarithmically with the intensity of the sensory signal. $\Delta H$, the change in $H$ that may occur with a change in signal intensity with time held constant, or with a change in time when signal intensity was held constant, enabled us to derive three psychophysical laws that had previously been known as empirical equations.

It is significant that the proposed $H$-function operates in the manner of a thermodynamic function. That is, it tells us what will be observed in the physical world without telling us the mechanism by which it will come to be. So, for example, the $H$-function can live in comfort with any mechanical/chemical system such as neural networks or cascades of biochemical reactions which can describe the operational details of sensory function. The thermodynamics of locomotive function works hand in hand with the principle of expansion of heated gases.

The second part of this work dealt with my attempt to bypass the $H$-function described above by appealing directly to physical entropy. That is, can we write $\Delta F = \kappa \Delta S$, where $S$ is physical entropy? It was demonstrated that this substitution does work, at least for one, restricted sensory system.

## 6.2. *Quantum Sufficiat*

Let us conclude with a nearly-serious description of perception on a minimalist scale. Imagine, purely for heuristic purposes, that a receptor exists capable of sensing only a single particle, using a reference signal of only a single quantum (Norwich and Wong [14]). The expression for $H$ with $m = 1$ becomes simply

$$H = \frac{1}{2} \ln(1 + \sigma_S^2 / \sigma_R^2) .$$

Expressing this equation in words,

$$H = \frac{1}{2} \ln\left(1 + \frac{\Delta(\textit{signal intensity})^2}{\Delta(\textit{reference signal intensity})^2}\right)$$

where the operator $\Delta$ signifies "the uncertainty in".

Ascribe to the signal particle the de Broglie wavelength, $\lambda$, so that

$$\lambda p = \mathrm{h}$$

where $p$ is momentum and h is Planck's constant. Now apply Heisenberg's Uncertainty Principle to the reference quantum, so that

$$\Delta E_{\text{reference}} \cdot \Delta t \geq h/(2\pi) \ .$$

During the interval $\Delta t$, the signal quantum will be detected, so the two processes will occur nearly simultaneously.

Dividing the latter two equations,

$$\frac{\lambda p/\Delta t}{\Delta E(reference)} \leq 2\pi$$

The incident particle loses its momentum, $p$, over the time interval $\Delta t$, the transferred momentum, $\Delta p$, exerting an impulsive force, $F$. The force, $F$, acts over a distance, $\Delta x$, the width of the wave packet defining the incident particle. It can be shown that $\Delta x$ may be expressed as an uncertainty of the order of magnitude of $\lambda$, so that $\lambda p/\Delta t = F \cdot \Delta x$ is a measure of $\Delta E(\text{signal})$. Hence,

$$\frac{\lambda p/\Delta t}{\Delta E(reference)} \sim \frac{\Delta E(signal)}{\Delta E(reference)} \leq 2\pi$$

Now, square both sides, add 1 to each side, take logs and multiply by $\frac{1}{2}$:

$$H = \frac{1}{2}\ln\left(1 + \frac{\Delta(signal\ intensity)^2}{\Delta(reference\ signal\ intensity)^2}\right) \leq \frac{1}{2}\ln(1+4\pi^2) = 1.85$$

So the perceiver of one quantum of energy (say one photon) cannot transmit to the perceiver more than 1.85 natural units (2.67 bits) of information. However, we recall from section 4.2 (footnote) that $1.85 = \ln 6.36$, which is, effectively, Miller's magical number — effectively the maximum information transmissible for a given "one-dimensional" signal. Therefore, perhaps this limitation takes origin directly from quantum physics, and governs all perceiving life forms.

**Acknowledgements**

I am most grateful to Brige Chugh for his critical comments on the manuscript. Many of these studies have been supported by the Natural Sciences and Engineering Research Council of Canada.

## References


1. K.H. Norwich, *Information, Sensation and Perception*, Academic Press, San Diego (1993).
   See also http://www.biopsychology.org/norwich/isp/isp.htm.
2. A. Moles, *Information Theory and Esthetic Perception*, pp. 22–23, University of Illinois Press, Urbana (1966).
3. J.M. Medina, *Phys. Rev. E* **79**, 011902 (2009).
4. W. Wong and S. Figueiredo, On the role of information and uncertainty in auditory thresholds. *Proceedings of the 2002 International Conference on Auditory Display, Kyoto, Japan, July 2–5* (2002).
5. R.R. Riesz, *Phys. Rev., Series 2*, **31**, 867–875 (1928).
6. F. Lemberger, F. *Plügers Archiv für die gesammte Physiologie des Menschen und Tiere* **123**, 293–311 (1908).
7. M. Wojtczak and N.F. Viemeister, *J. Acoust. Soc. Am.* **106**, 1917–1924, (1999).
8. H. Piéron, *The Sensations*, Yale University Press, New Haven (1952).
9. H. Doma and P.E. Hallett, *Vision Res.* **28**, 899–913 (1988).
10. E.A. Jackson, *Equilibrium Statistical* Mechanics, p. 99. Prentice-Hall, Englewood Cliffs, N.J. (1968).
11. K.H. Norwich, *Chem. Senses* **26**, 1015–1022 (2001).
12. E. Von Skramlik, *Handbuch der Physiologie der Niederen Sinne*, pp. 458–460. Georg Thieme, Leipzig (1926).
13. C. Pfaffmann, L.M. Bartoshuk and D.H. McBurney, In Beidler, L.M. (ed.) Handbook of Sensory Physiology: Vol IV, Chemical Senses 2: Taste, pp. 85–86. Springer, New York (1971).
14. K.H. Norwich and W. Wong, *Annales de la Fondation Louis de Broglie* **22**, 161–168, 199.

# THE COLLECTIVE BRAIN

ENZO TAGLIAZUCCHI

*Departamento de Física, Facultad de Ciencias Exactas y Naturales,*
*Universidad de Buenos Aires, Argentina*
*E-mail: nztglzcch@gmail.com*

DANTE R. CHIALVO

*Consejo Nacional de Investigaciones Científicas y Tecnológicas (CONICET),*
*Buenos Aires, Argentina*
*Facultad de Ciencias Médicas, Universidad Nacional de Rosario, Rosario, Argentina*
*David Geffen School of Medicine, UCLA, Los Angeles, CA, USA*
*E-mail: dchialvo@ucla.edu*
*www.chialvo.net*

The unique dynamical features of the critical state can endow the brain with properties which are fundamental for adaptive behavior. This proposal, put forward with Per Bak several years ago, is now supported by a wide body of empirical evidence at different scales demonstrating that the spatiotemporal brain dynamics exhibits key signatures of critical dynamics previously recognized in other complex systems. The rationale behind this program is discussed in these notes, followed by an account of the most recent results, together with a discussion of the physiological significance of these ideas.

*Keywords*: criticality; brain dynamics; phase transition; complexity.

## 1. Introduction

Throughout the years, brain theory incarnated in various forms following contemporaneous technology. As expressed by Braitenberg:[12] *...Fascinating aspects of the latest technology are always creeping into science and being turned into subconscious motives for theory. For brain science, in the nineteenth century it was optics, with its idea of projection, and in the twentieth century it was radio engineering, with its logical circuits...*". In the nineties, the fundamental concepts behind the physics of complex systems, motivated us to work on ideas that now seem almost obvious: (1) the mind is a collective property emerging from the interaction of billions of agents; (2) animate behavior (human or otherwise) is inherently complex; (3) complexity and criticality are inseparable concepts. These points were not chosen arbitrarily, but derived, as discussed at length here, from considering the dynamics of systems near the critical point of a order-disorder phase transition. Simply put, this view considered the brain as *just another* dynamical system at criticality, knowingly non unique.

The brain seen as a dynamical object as discussed in these notes is grounded on accessible empirical evidence: as brain activity unfolds in time at all spatial scales, different patterns evolve. Proper measurements provide quantitative information regarding these patterns (for example, neuron spike trains, local field potentials, metabolic signals, behavioral measures, etc). The question is whether is it possible to explain all these results from a single fundamental principle. And, in case the answer is affirmative, what does this unified explanation of brain activity implies about goal oriented behavior? We will submit that, to a large extent, the problem of the dynamical regime at which the brain operates it is already solved in the context of critical phenomena and phase transitions. Indeed several fundamental aspects of brain phenomenology have an intriguing counterpart with dynamics seen in other systems when posed at the edge of a second order phase transition.

The paper is organized as follows: first, emergent complex phenomena and the intimate connection between criticality and complexity will be described. This will be in Section II, which reviews earlier work on ant's swarm, a metaphorical mind in itself, which is mathematically very close to nowadays models of collective decision-making. The large scale analysis of brain dynamics will be introduced in Section III, where the experimental results indicating criticality are discussed in detail. Section IV provides examples of how complexity and criticality manifest themselves in a relatively smaller scale (scale of a few millions of neurons). As mentioned above, the statistical properties of the brain dynamics must be reflected in the dynamics of behavior. This is treated in Section V where recent reports seems to show such correspondence. Section VI, dwells into the evolutionary argument by which the brain and the animate behavior must be critical anytime an organism needs to survive and evolve in an environment which, by thermodynamic reasons, is also critical. Section VII is dedicated to discuss the tendency in the field to consider equilibrium models driven by external noise to accommodate the empirically observed fluctuations. Section VIII closes the paper with a prospective of some relevant issues to pursue further.

## 2. Emergent Complex Dynamics is always Critical

Emergence refers to the observation of dynamics that is not expected from the systems equations of motion and, almost by (circular) definition, is exhibited by complex systems. As discussed at length elsewhere,[3,15,17,19,37,47,56] three features are present in complex systems: (I) they are *large* conglomerate of *interacting* agents, (II) each agent own dynamics exhibits some degree of *nonlinearity* and (III) energy enters the system.[a] These three components are necessary for a system to be able to exhibit, at some point, emergent behavior. It is well established that a number of isolated linear elements cannot produce unexpected behavior (mathematically, this is the case in which all motion can be formally predicted).

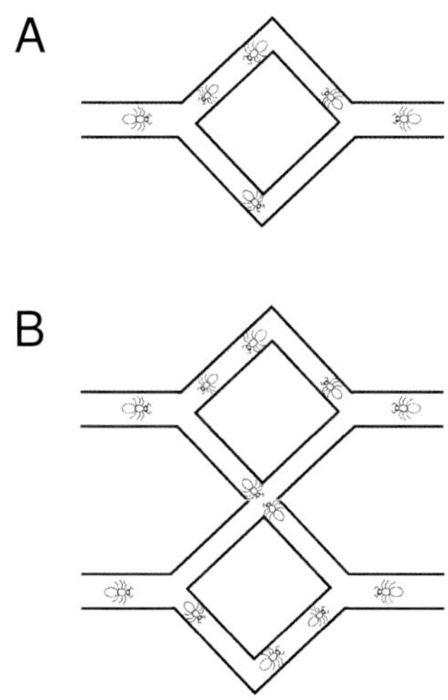


Fig. 1. A: In typical binary bridge experiments, ants are allowed to walk over bridges connecting two or more areas with different concentration of pheromones. The dynamics of decision-making changes depending on the pheromone lying in the surface and on the ant's density. B: Gedanken experiment: the beginning of a network constructed by connecting binary bridges.

[a]As an example, think that the agents are humans who are driven by food, sunlight and other energy sources, they can form families and communities. After many generations, eventually some political structure arises. Whatever the type of structure that emerges is unlikely to appear in the absence of any of the three elements emphasized above.

An inspiring example of emergence of complexity is the dynamic of swarms,[33–35] which we used in the past as a toy model to understand how cognition could arise in neural networks.[21,42] Of course, the case for social insects representing a paradigm of cooperative dynamics was extensively discussed before.[59] The uncertainty was centered at what collective property of foraging ants allows the emergence of trails connecting the nest with the food sources, forming structures spanning sizes several order of magnitude larger than any of the individual's temporal or spatial scales. To clarify that, Millonas introduced[33] a spatially extended model of what he called "protoswarm". His objective was to understand the microscopic mechanism by which relatively unsophisticated ants can build and maintain these very large macroscopic structures.[b]

In the swarm model, there are two variables of interest, the organisms (in large number) and a field, representing the spacial concentration of a scent (such as pheromone). As in real ants, the model's organisms are influenced in their actions by the scent field and in turn they are able to modify it by depositing a small amount of scent in each step. The scent is slowly evaporating as well. The organisms only interact through the scent's field. The model is inspired in the behavior of real ants, in which they are exposed to bridges connecting two or more areas where the ants move, feed, explore, etc. Eventually they will discover and cross one of the bridges. As it is illustrated in Figure 1A they will come to a junctions where they have to choose again a new branch, and continue moving. Since ants both lay and follow scent as they walk, the flow of ants on the bridges typically changes as time passes. In the example illustrated in Figure 1A, after a while most of the traffic will eventually concentrate on one of the two branches. The collective switch to one branch is the emergent behavior, something that can be understood intuitively on the basis of the positive feedback between scent following, traffic, and scent laying. What is less obvious is linking these ideas with the microscopic rules governing the dynamics of a single ant. Numerous mathematical models and computer simulations were able to capture the ants behavior in the bridge experiment.[8,22,38]

The bridge experiment results can be used as building blocks to more sophisticated settings, such as ants freely exploring an extended arena. The insight comes from a rather clever way that Millonas's model discretized space. The model can be viewed, for descriptive purposes, as a network constructed by connecting each point of a square lattice to its eight nearest

[b]Most readers will recognize the similarities with brain's phenomenology.

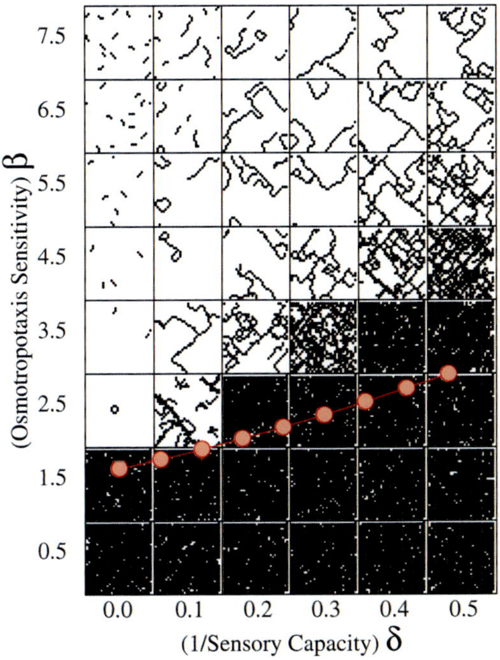


Fig. 2. Disordered, complex and ordered patterns can be observed as asymptotic dynamic in synthetic swarms for different values of gain ($\beta$) and dynamic range ($\delta$) in the ants' sensory system. For each parameter $\beta$ and $\delta$ value indicated on the axis the dots in each of the 48 squares represent a snapshot of the ants' positions at the last ten iterations (after 5000 time steps starting from a random initial distribution of agents). The circles joined by a line are the values for the phase transition predicted by the theory.[33–35] There are three distinct behaviors: one region (below the line) where ants trajectories are random, eventually covering the entire arena; a second dynamics (near and above the line) where lines of traffic over complex paths appear, and yet another patterns with clusters of immobile ants (far above the line). The arena is a 32x32 lattice and the other parameters as in Refs. 21 and 42. Replotted from Ref. 18.

neighbors, as in the cartoon of Figure 1B. Thus, at each step each ant makes a decision to choose one of eight bridges; and deposits a fixed amount of pheromone as it walks. The decision is based on the amount of scent at each of the eight locations. The ants' sensory apparatus embedded in a physiological response function was modeled following biological realism, having two parameters: one which could be considered analogous to gain and the other the inverse of sensory capacity (or dynamic range). The plot in Figure 2 condenses the results from the numerical simulations with different values for the physiological response function. At each combination of the explored $\beta$ and $\delta$ values there is a square plot which depicts the locations of each ant at the last ten steps of the simulation. It can be seen that ants converge to different behaviors depending on the parameter' values. For values of both small gain and small dynamic range ants execute a random path, resulting in the plots fully covered, as in the right bottom corner. Thus, because of the low sensitivity ants are just making random choices at each juncture. For large enough gain (top left corner), ants senses saturate resulting in clusters of immobile ants in the same attracting spot. It is between these two states, one disordered and the other frozen, that the swarm can organize and maintain large structures of traffic flow as those seen in nature.

The stability analysis in Ref. 33 makes straightforward to understand the transition between the disordered walks and the complex structures of trails (line and circles in Figure 2). Intuition already suggests that the model's $\beta$,$\delta$ values at which the order-disorder transition happens must depend on the number of ants able to reinforce the scent field. In Figure 3

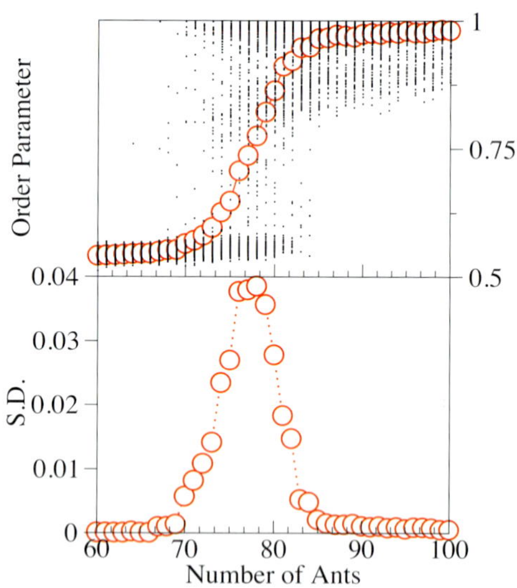


Fig. 3. Collective swarm behavior near the phase transition. Collective order (top) increases as a function of increasing density of ants (i.e., the pheromone field is more frequently renewed). Note the generic increase in the amplitude of the fluctuations of the order parameter (bottom) near the critical point. Order parameter values from individual runs are plotted using dots and circles for the average. S.D: Standard deviation of the data in the top panel. Replotted from Ref. 18.

results from several hundreds runs with fixed parameters and increasing density of ants are shown. The degree of collective order was evaluated by the proportion of ants "walking" on lattice points having above average scent concentration (see further details in Refs. 21 and 42). The top panel shows a plot of the results where the dots indicate the outcome of each individual run (with different initial conditions) and circles the average of all runs. For low ants' density the expected random behavior is observed, with equal likelihood for ants to be in or out of a high scent field. For increasing number of ants the swarm suddenly starts to order until it reaches the point in which the majority of the ants are walking on a field with scent concentration larger than the average. Notice that, as the density approaches the transition, the amplitude of the order parameter's fluctuation increases (see bottom panel of Figure 3), a divergence which is generic for criticality. Thus, at the transition, large trial-to-trial variability is found for repeating realizations of the same numerical experiment. The most significant point here is to note that the best performance for the swarm, namely where long trails are formed (see Figure 2) corresponds also to the conditions for maximum variability. Thus maximum variability and performance coexist.[c]

These results were the first to show the simplest (local, memoryless, homogeneous and isotropic) model which leads to trail forming, where the formation of trails and networks of ant traffic is not imposed by any special boundary conditions, lattice topology, or additional behavioral rules. The required behavioral elements are stochastic, nonlinear response of an ant to the scent, and a directional bias. There are other relevant properties, discussed in detail elsewhere,[21,42] that arise *only* at the region of ordered line of traffic, including the ability to reconstitute trails and amplify weak traces of scent, in analogy to memory traces. The conclusions important to these notes are that simple local rules allow the emergence of complex self organized patterns, which extend in a non local way beyond the scale of a single agent and which appear suddenly in a dynamical regime between random and uniform behavior, the "critical state".

The remaining of these notes will show evidence of an analogous behavior when neurons interact nonlinearly in the human brain, giving rise to the emergence of complex non-local patterns, as well as the functional and behavioral consequences of these patterns.

[c]The brain correlate for this divergence will be a large variance in some order parameter (extracted from the dynamics) coexisting with optimal response to the same input, for instance.

## 3. The Collective Large-scale Brain Dynamics

To visualize the collective activity of the individual constituents of the human brain (neurons) is harder, of course, than to contemplate the ants of the previous sections. For this purpose different methodologies have been developed which are capable to register this activity at different timescales and with different limitations. Large scale activity (resulting from the averaging of thousands of neurons) can be measured with excellent spatial and good temporal resolution using fMRI (functional magnetic resonance imaging). This non-invasive technique allows indirect measurements due to metabolic and oxygenation changes correlated with synaptic activity, finally encoded in the BOLD (Blood Oxygen Level Dependent) signal.

Patterns of global activity can be analyzed studying the relationship between the behavior of different members of the collective, in this case, regions of cortical tissue on the millimeter scale. Functional interactions between these areas can be studied comparing their BOLD signal timecourses. These functional interactions are not only present when the brain engages in a task but also during rest (i.e. spontaneous activity, in contrast to evoked activity). During the remainder of this notes we will focus mostly on resting state brain dynamics. These interactions can be represented using the concept of a network or a graph.[15,24,45,48,49,57] A graph consists of nodes connected by links, in this case the nodes are voxels (the smallest cubic regions that the spatial resolution of the method can resolve) and links represent coordinated activity between these two linked voxels.

Different measures of coordination between the time course of the BOLD signal may be employed, the simplest being linear correlation. Functional connectivity networks can be constructed computing the linear correlation between BOLD signal pairwise for all voxels and introducing a link between voxels if they are correlated beyond an arbitrary threshold (for a diagram of the procedure see Figure 4). For a wide and reasonable range of thresholds, functional brain networks thus constructed are scale free, this is, the degree histogram of the voxels follows a power law of exponent approximately 2 (where degree is defined as number of neighbors in the network).[24,57] Also, functional connectivity networks have the small world property[1,6,7,50] (the mean number of links that must be crossed to connect a given pair of nodes is small relative to the local structure of the network) and are assortative (connectivity is stronger between nodes of similar degree). These properties are shared with a large number of networks constructed from different biological and social systems.[48,49] They have strong implications regarding

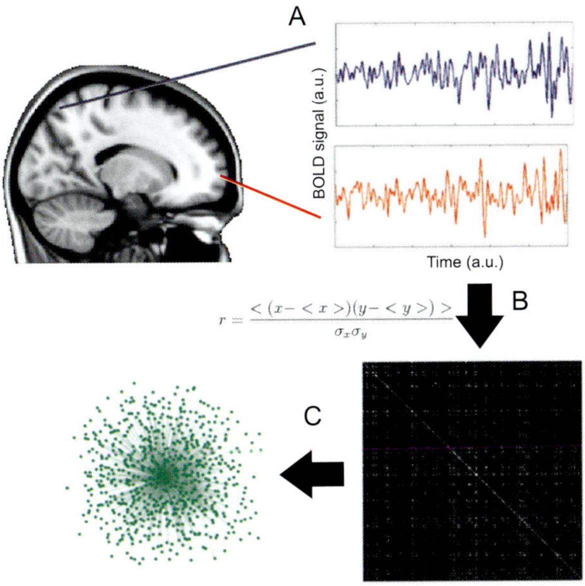


Fig. 4. Procedure to extract brain functional connectivity networks. A: BOLD signal is extracted for every pair of voxels (in this case, voxels in the precuneus and middle frontal gyrus). B: Linear correlation (r) is computed between each pair of signals, leading to a correlation matrix. C: Correlations are thresholded at a given arbitrary value, a link is established between two nodes (or voxels) if the correlation of their respective BOLD signals exceeds the threshold, leading to a graph or network.

information transfer, stability and general robustness of the networks. Together, the scale free and small world properties imply the presence of long range (non-local) functional interactions between brain regions. Most important to these notes, functional connectivity networks constructed from time-courses extracted from a paradigmatic example of a system with critical dynamics (Ising model for ferromagnetic materials) are virtually indistinguishable from brain functional connectivity networks.[26] This striking similarity between both networks is not imposed in the model: it arises as a very general consequence of critical dynamics, thus giving an explanation to the topological properties of brain functional connectivity networks solely based on the concept of criticality.

Functional connectivity networks are a useful global description of the brain, albeit a static one. They can be regarded as the "skeletons" of the dynamical systems, over which patterns of activity unfold over time. Deeper insight is gained from dynamical analysis (temporal and spatiotemporal)

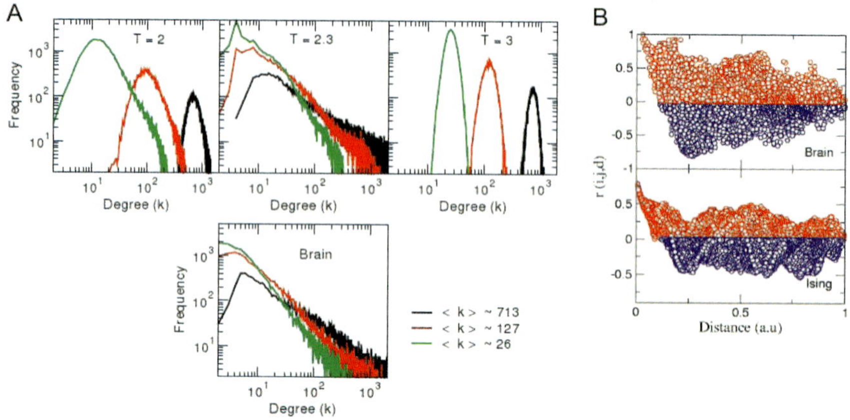


Fig. 5. A: Degree distribution of functional connectivity networks extracted from the Ising model at different temperatures and different mean degree ($\langle k \rangle$) (up) and degree distributions of brain functional connectivity networks (down) B: Correlation vs. distance to a given node for brain functional connectivity networks (up) and for Ising model networks (down). Adapted from Ref. 26.

of large scale brain activity. Imaging experiments have gathered evidence showing that (even in the absence of explicit perception and cognitive performance) this activity displays complex behavior. In the purely temporal domain, it is not homogeneous nor periodic (as EEG, MEG, fMRI, etc, recordings indicate) neither there is a representative or predominant frequency, as shown by the 1/f decay of spectral density.

In the spatiotemporal domain, fMRI whole brain recordings reveal how activity explores complex patterns with long range correlations and anticorrelations.[27,28] Strikingly, these patterns can be reduced to a small number of prototypical ones using a mathematical technique termed Independent Component Analysis (ICA).[9,10] Careful examination shows that these structures appear and disappear over time in the fashion of "passing clouds" that last only a few seconds. It has been shown that the spatial maps of these resting state independent components (also called resting state networks or RSNs) can be associated with different cognitive functions and are strongly correspondent with activation maps obtained during task performance.[46]

This correspondence supports a dynamical view of resting state, in which activity continuously explores patterns (which are "metastable states") only to get temporarily locked in one of them as a task is executed

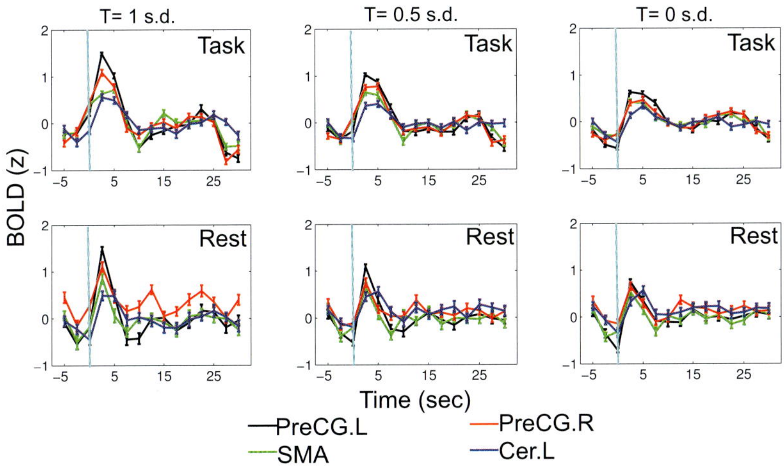


Fig. 6. Averages of the BOLD signal triggered by increments of activity (beyond 0.5 S.D. from the mean) in left primary motor cortex (PreCG.L) during right finger tapping (up) and resting state (down). Whenever the signal at PreCG.L exceeds the threshold, 30 sec. of the signal are extracted from right primary motor cortex (PreCG.R), supplementary motor area (SMA) and left cerebellum (Cer.L) and averaged. Notice the similarity between task and rest conditions, not only in shape but also in BOLD intensity. Replotted from Ref. 52.

or a sensory system stimulated. Not surprisingly, it is when operating at the critical point that a dynamical system becomes most efficient in the *flexible* exploration of these metastable states. This dynamical view of resting state need to be exploited given the strong support[52] from recent experiments involving simple motor tasks and resting state. Briefly, it has been shown that average BOLD signals associated with spontaneous co-activations of key motor cortical areas strongly resemble co-activations during evoked activity, not only in location and temporal shape but also in intensity. In other words, spontaneous activations are completely analogous to evoked activations, except that the first ones are transient and their timing is uncertain. This correspondence is shown in Figure 6 for regions in the primary motor cortex, supplementary motor area and cerebellum, during finger tapping and rest. The continuous exploration of these patterns during resting state may have strong neurophysiological relevance, yet to be fully understood.

It must be noted that while resting-state and spontaneous brain activity seems an ill defined concept, the bulk of results described in this last paragraph has not only been observed in conscious behaving humans but also during sleep and anesthesia. A great consistency is observed across subjects: approximately 10 minutes of fMRI measurements from a single subject are enough to reveal the main resting state networks (as shown in Figure 8).

This dynamical scenario is highly consistent with the brain operating in a critical regime. A final observation regarding whole brain measurements is that the lack of measurement scale implied by critical dynamics is also manifest in the invariance of the two-point correlation function of the BOLD signal after a normalization procedure (spatial coarse graining) is repetitively carried out.[23] Additional evidence is derived from the computation of the correlation length of the BOLD signal fluctuations, which are known to diverge near the critical point. Figure 7 shows experimental results obtained from human fMRI at rest indicating that correlation length is not constant but diverges with the size of the cluster considered.[20] This demonstration is a landmark of critical phenomena, together with previous observations[23,32] of scale invariance at large scale add weigth to the contention of criticality as the brain's dynamical state.

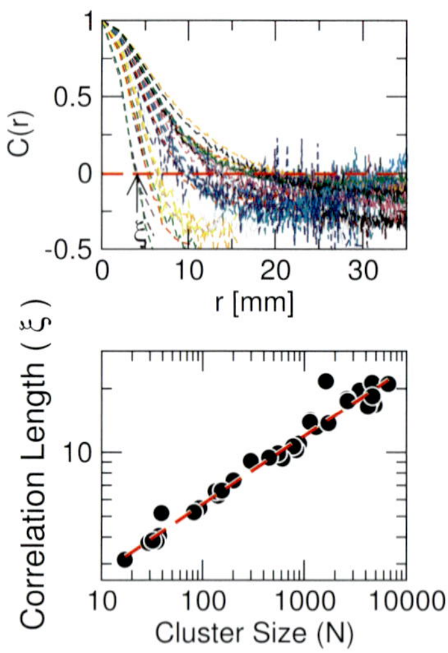


Fig. 7. Top panel: increase of correlation length with cluster size. Each line shows the cross-correlation of BOLD activity fluctuations as a function of distance averaged over all time series of thirty five clusters (fragments of resting state networks). Bottom panel: Functional dependence of the correlation length (crossing of each curve of top panel through zero). The correlation length grows linearly with the cluster diameter. Thus, for example, the BOLD activity of two voxels spaced 4 mm apart on a cluster composed by 20 voxels, are as strongly correlated as two points 10 mm apart on a cluster composed by 1000 voxels. Replotted from Ref. 20.

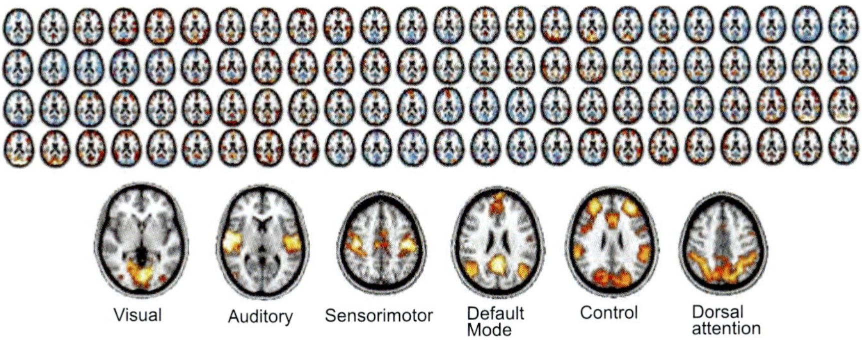


Fig. 8. Up: Approximately 10 minutes of resting state brain activity from a single subject (red-yellow for BOLD signal above the mean, blue-purple for BOLD signal below the mean). Images are separated by 2.5 s. Down: Main resting state networks obtained through seed based correlation from data shown above. Replotted from Ref. 19.

## 4. Neuronal Avalanching in Small Scale is Critical

Historically, the first direct demonstration of collective critical dynamics was the observation that a cultured slice of neurons supported "avalanches", this is, intermittent bursts of activity that spread up to the whole system.[11,41] This dynamical state is halfway between highly synchronized oscillatory activity and disordered noise. Even though pairwise correlations are low, and with highest probably only a small number of neurons fire in synchronized fashion, occasionally activity spreads in the form reminiscent of an avalanche in Per Bak's sandpile model. Power laws are observed for event size (number of electrodes registering electrical activity) with exponents agreeing with those of a critical branching process. These results have been replicated in other experimental settings, including cortical measurements in awake, behaving monkeys[39] and rats.[43]

Modeling efforts to understand the precise neuronal cause of avalanches are currently being made. Certain models are able to replicate these scale free densities while explicitly avoiding criticality, however, recent numerical results show that their predicted exponents are in the wrong order of magnitude as those experimentally observed. Also, even if avalanches are a dynamical process, explanations of scale free behavior can be based on the underlying connectivity of the neurons. However, directly incorporating long range correlations in the structural connectivity of the collectivity solves one problem (integration) but succumbs to a different one (segregation). Critical dynamics is, thus, a natural constraint to incorpo-

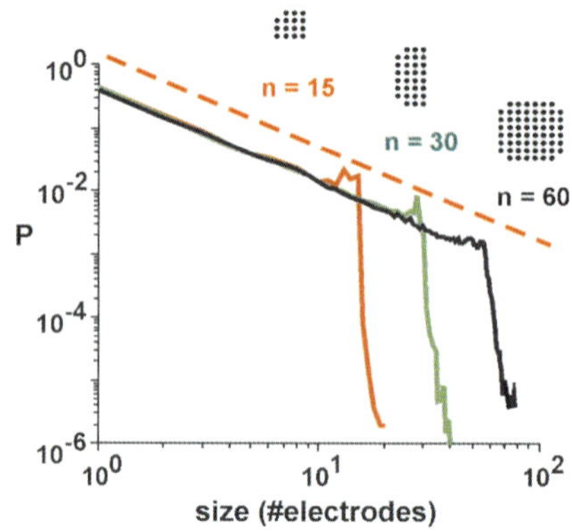


Fig. 9. Size distribution for avalanches in mature cortical cultured networks and different electrode array sizes (15,30,60 electrodes). Power laws with different cutoffs arise for each number of electrodes (finite size scaling). Measurements comprise 70 hours of data recording. Replotted from Ref. 15.

rate into a model of neuronal avalanches. Models of neurons with activity dependent synaptic coupling undergoing self organized criticality reproduce statistics of neuronal avalanches, even with trivial (random) connectivity of its elements.

Finally, even though the most salient features of experimental findings regarding neuronal avalanches are power laws probability densities for size and event durations, there are other predictions to be fulfilled if the underlying dynamics are in fact critical. In short, these are: I) time scales separation between the dynamics of the triggering event of the avalanche and the avalanche itself, II) stationary avalanche size statistics regardless of avalanching rate fluctuations (excluding non-homogeneous Poisson processes), III) Omori's law for earthquakes must apply to avalanche probabilities after and before main events, IV) average size of avalanches following a main avalanche decays as an inverse power law, V) avalanches spreads spatially on a fractal. Recently, it has been reported[40] that experimental data in fact supports these predictions, thus narrowing the search of models of avalanches to those undergoing critical dynamics.

## 5. Psychophysics and Behavior

As exposed in the previous sections, the hypothesis that collective brain dynamics operates at a critical point has received plenty experimental support. This evidence spans a wide range of scales both in time and space and leads in a natural way to a discussion of consequences at a behavioral level. Simply put: if brain activity is critical, signatures of criticality must be as well observed in behavior and perception.

An old unsolved problem of psychophysics relates to the need of a very ample dynamic range of neuronal responses, spanning several orders of magnitude.[16] For example, visual perception is known to adapt from very dimly lit environments (such as caves, forests, etc) to landscapes of high luminosity; however, the dynamic range of an isolated neuron (spanning at most a single order of magnitude) is too limited to do so. The key to solve this problem resides in the consideration of a connected network of excitatory neurons which has an amplification factor of one. In this model, neurons operate in a critical point, in contrast to a subcritical regime in which sensory inputs quickly die, or a supercritical regime, in which any input spreads in an explosive fashion throughout the network.[31] These three possibilities are exemplified in Figure 10. Thus, the fact that a system at the critical point has the greatest dynamical range, is now the key to solve the dilemma of human perception optimized over several orders of magnitude.

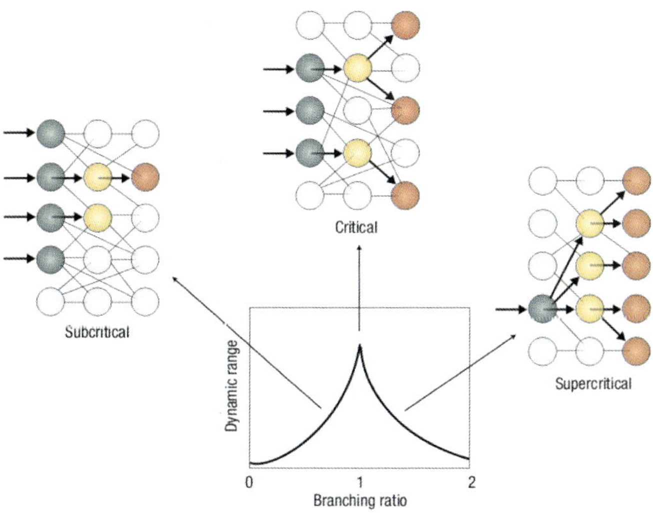


Fig. 10. The three dynamical regimes of a sensory network model: subcritical regime (amplification or branching ratio less than one), critical (amplification equal to one), and supercritical (amplification greater than one). In critical networks, activity levels are conserved, and the dynamical range of responses is maximized. Redrawn from Ref. 16.

On the other hand, animal behavior has been shown to reflect the critical dynamics of brain activity. Recordings of spontaneous motion activity of rats during large periods of time reflects scale free distribution for length of movement times and pauses.[2] This result, observed in an animal model and controlled laboratory conditions, has similarities with reports on human activities such as writing of letters, emails, web browsing and motion initiation. Together, these results suggest the presence of behavioral signatures of critical dynamics in animals and human beings.

## 6. An Evolutionary Perspective

The ingredients for complex emergent behavior presented in the first section of these notes (a large number of energy-driven nonlinearly interacting elements) are not only present in the human brain, but are also ubiquitous in the whole natural world.[3,4,13,37] Humans and other animals navigate in this complex critical world, which as such offers constantly surprises and unexpected events. One can argue that in a subcritical and uniform world a brain is a superfluous organ since there is nothing to learn, on the other hand, an ever-changing supercritical world offers no regularities to be learned.[5,14] In the middle of these two extreme situations there is a real need for a learning device. The coupling between a critical world and a critical brain is best exemplified in the playing habits of children. Young children become quickly bored of inanimate toys (even the addition of sophisticated motion mechanisms does not dramatically improve this situation), however, they obtain unending fun from animal pets for long periods of time. The reason behind is that animate behavior is a mixture of order and surprise in the same sense than the spatiotemporal patterns of a system at the critical point are composed of a blend of order and disorder. Thus, the need of a brain stems from the fact that the world around is critical, thus, in turn evolutionary pressures constraint brains to operate in a critical regime.

Brains must not only be capable of mere learning: the ability to forget is central to adaptive behavior. In a sub-critical brain state of highly correlated activity, memories would be "frozen" and ever present. On the other hand, a supercritical brain would have its activity patterns changing wildly and constantly in time, resulting in the inability to retain any memories. The conclusion is that, in order to adapt to a complex world that constantly presents us with novelties, evolution has created brains and forced them operate at the critical point.

## 7. Noise or Critical Fluctuations? Equilibrium vs Non-equilibrium

As discussed in Section II, criticality is a universal dynamics exhibited by a large class of systems. Consequently it offers a unified theoretical framework to explain a myriad of apparently unrelated observations. We will briefly provide an illustration of this possibility, discussing work recently reported, where the current theoretical understanding can be fruitfully applied.

In a series of papers[29,36] the variance and mean-based spatial fMRI patterns obtained in subjects of different ages were compared, concluding that the descriptive value of the variance was greater than the averages. The authors suggested that *"examination of BOLD signal variability may reveal a host of novel brain-related effects not previously considered in neuroimaging research"*. Additional work (by the same authors) went even further, revealing that BOLD signal variability is greater at younger age, confirming that *"younger, faster, and more consistent performers exhibited significantly higher brain variability across tasks, and showed greater variability-based regional differentiation compared to older, poorer performing adults"*.[30]

The origin of this variability was explained[36] by suggesting that: *"The interplay between local and global dynamics governs the spatiotemporal configuration of the brain's functional architecture and keeps the system in a high-energy state, at the 'edge of instability' among a number of different states and configurations."*, while the switch between such configurations was suggested to happen by the action of noise originated in the following manner: *"Intrinsic neural noise — stochastic fluctuations in information transfer caused by imprecise timing of cellular processes — serves to nudge the system from one state to another and thus confers the capacity to make fluid and adaptive transitions between different states and reconfigure either spontaneously or in response to external (task) demand."*

Besides the obvious merits of calling attention to the variability, the crucial question here should be how much real understanding was gained by finding that the variability was more informative than the average.[d] But lets postpone the answer to this question until we consider yet another example.

[d]The first and second moments are usually linked by some functional relation, thus it is not completely surprising that averages and variance share predictability.

The second example is concerned with a very interesting book[44] dedicated recently to the topic of noise in the brain, which in its very first five lines reads: *"The relatively random spiking times of individual neurons produce a source of noise in the brain. The aim of this book is to consider the effects of this and other noise on brain processing. We show that in cortical networks this noise is an advantage, for it leads to probabilistic behavior that is advantageous in decision making, by preventing deadlock, and is important in signal detectability"*. This is not an isolated quote taken out of the proper context. The authors enumerate a three "reasons why the brain is inherently noisy and stochastic"[44] including sensory noise arising external to the brain, cellular noise due stochastic opening and closing of ion channels, and synaptic noise.

These two examples are in line with nearly all detailed models of neuronal function which require ad-hoc noise inputs to function properly. Simply put, without noise the system is at equilibrium, stuck on a stable state. All these analysis overlook a crucial question: where does this (fine tuned or not) noise comes from? If it is generated by specific systems within the brain, then experimental evidence ("a noisy center"?) should be found pointing to where these systems are and how they work. Of course, sarcasms aside, no such center would be found. Instead, ad-hoc noisy driving is not required by the theory of critical brain dynamics, since noise is self-generated by the collective dynamics which spontaneously fluctuates near the critical point as was shown in Figure 3. It is well known, that large fluctuations are inherent to a system approaching the critical point,[3] and in that sense "variability" is perfectly understood as part of the critical dynamics, and not a force external to the system. In our opinion, the often conjectured reasons for the noise presence, assigning a purpose to it,[e] are *teleological* dead-ends that postpone true understanding.

The fact is that the correct interpretation of the noise is a much deeper issue, which reflects the general tendency to consider only equilibrium models. These models achieve a better agreement with empirical observations only with the addition of *external* noise (see for instance Chapter 12 in[48] or work compiled by Steyn–Ross[51] recently). This deformation is not exclusive of neuroscience and its consequences are far reaching. For example, similar views in other complex systems include the explanation of episodes with very large number of species going extinct throughout earth history by adding external noise, such that for each extinction an out of space event

[e] "the noise serves to...,"

(meteorite) hits the earth. In contrast, non-equilibrium models[3,37] show that similar exceptionally rare and large extinctions (and their mirror situations, namely speciations) can be inherent of the co-evolutionary species dynamics, and that large extinctions could have happened in absence of meteorites. In these type of models, variability and disparate fluctuations are (at criticality) typical. Another example belongs to macro-economics, as the late Per Bak hammered many times in his writings[3] and lectures, criticizing models in which each market crash would have required an event that drove the shares to plunge. In analogy, historians will assure us that in Sarajevo on the 28th of June of 1914 a member of the Black Hand, a Serbian nationalist secret society, assassinated Archduke Franz Ferdinand, heir to the Austro-Hungarian throne. The equilibrium view will identify this single event (noise?) as responsible to trigger the mobilization of 65 millions soldiers, 7 millions missing, 8 millions killed and 21 millions wounded. In contrast, the non-equilibrium approach will assert that the First War World (and other infrequently large conflicts) will eventually happen in the absence of an isolated gunman.

Thus, coming back to the brain, where its main dynamics are fluctuations, it need to be recognized that having to add fluctuations (noise) to the model equations to get fluctuating dynamics is a serious limitation. It seems reasonable that future work consider that the differences between modeling a "noisy brain" or a "critical brain" are significant and relevant, because the gap separating these two views is conceptually as large as for physics is the difference between equilibrium and non-equilibriums dynamics.

## 8. Outlook

The study of collective dynamics and critical phenomena was pioneered by physicists during the past century, in order to solve problems related with material science and solid state physics. It is now clear that the theoretical framework used for these problems is very general and can be applied to understand an impressive range of phenomenology from many disciplines. Emergent dynamical properties of interacting agents (phase transitions, divergence of correlation length and fluctuations, maximum dynamical range, etc) are far from explicit in the individual equations of motion, thus, a reductionist approach will be (by its very definition) unable to explain this emergence.

Pure reductionism has in fact been abandoned in a large list of fields which study complex many particle systems, from economics and sociology to molecular biology. It is surprising that this revolution has only recently began to have an impact on neuroscience, a discipline which deals more than any other with a collective of interacting agents. Still today, a large number of published results stem from detailed experiments or simulations of neurons or networks of millions of neurons, including sometimes great biological detail but no considerations related with the statistical physics of collective phenomena in large systems. However, after the realization that theories of brain function must be compatible with the laws of collective behavior, a non-return point has been reached: a rapidly increasing number of reports being deal with theoretical and experimental evidence grounding the emergent critical dynamics hypothesis.[19]

Researchers now find that paradoxical situations and seemingly disconnected facts can be re conciliated within the critical brain dynamics theory, without need of complicated ad-hoc hypothesis. A good example of this is the integration/segregation dilemma,[53–55] a problem haunting neuroscientists since the beginnings of the discipline: how does the brain manages to integrate information from different modalities into a single decomposable scene, while making easy to segregate particular aspects of this scene at will? Or in other terms: how does brain activity from different regions coordinates without collapsing into a single block of uniform dynamics? A concrete example is the synchronous coexistence of multiple operational modules studied by Fingelkurts & Fingelkurts.[25] Many connectionistic approaches have been taken to this problem, implying that coordination is hardwired into anatomical connectivity. These proposed solutions are able to solve one problem (integration) but still fail to explain the ease of brain dynamics to segregate. Within the theory of critical brain dynamics, however, this situation arises naturally: even systems with trivial structural connectivity (for example, the first-neighbor connections of the Ising model) achieve (transient) long range correlations at the critical point.[26]

Any scientific theory needs to explain and integrate previous facts, solve apparently paradoxical situations, and make predictions amenable to experimental verification. Along these notes we have shown how these requirements are fulfill by the theory of emergent brain collective critical dynamics. The transfer of ideas from statistical physics to other disciplines is a young and exciting endeavor, even younger in neuroscience,[58] however exciting results have already been achieved. Only with the realization that brain is a collective (and henceforth must follow the physical laws that govern collectives) a deep understanding of brain and human behavior will be achieved.

## Acknowledgements

Work supported by NIH of USA (Grant NS58661) and by CONICET (Argentina). E.T. was supported by an Estímulo Fellowship from the University of Buenos Aires.

## References

1. S. Achard, R. Salvador, B. Whitcher, J. Suckling, E. Bullmore, Resilient, low-frequency, small-world human brain functional network with highly connected association cortical hubs. *J Neurosci.* **26**, 63 (2006).
2. C. Anteneodo & D.R. Chialvo, Unraveling the fluctuations of animal motor activity. *Chaos* **19**, 033123 (2009).
3. P. Bak, *How Nature works.* (Oxford University Press, Oxford, 1997).
4. P. Bak, C. Tang, K. Wiesenfeld, Self-organized criticality: an explanation of the 1/f noise. *Phys. Rev. Lett.* **59**, 381 (1987).
5. P. Bak & D.R. Chialvo, Adaptive learning by extremal dynamics and negative feedback. *Phys. Rev. E* **63**, 031912 (2001).
6. D.S. Bassett, A. Meyer-Lindenberg, S. Achard, T. Duke, E. Bullmore, Adaptive reconfiguration of fractal small-world human brain functional networks, *Proc. Natl. Acad. Sci. USA* **103**, 19518 (2006).
7. D.S. Bassett & E. Bullmore, Small-World Brain Networks *Neuroscientist* **12**, 512 (2006).
8. R. Beckers, J.L. Deneubourg, S. Goss, J.M. Pasteels, Collective decision making through food recruitment. *Insectes Soc.* **37**, 258 (1990).
9. C.F. Beckmann & S.M. Smith, Probabilistic independent component analysis for functional magnetic resonance imaging. *IEEE Trans. Med. Imag.* **23**, 137 (2004).
10. C.F. Beckmann, M. DeLuca, J.T. Devlin, S.M. Smith, Investigations into resting-state connectivity using independent component analysis. *Phil. Trans. R. Soc. B* **360**, 1001 (2005).
11. J.M. Beggs & D. Plenz, Neuronal avalanches in neocortical circuits, *J. Neuroscience* **23**, 11167 (2003).
12. V. Braitenberg, Some arguments for a theory of cell assemblies in the cerebral cortex. in *Neural connections, mental computation*, eds. L Nadel,L Cooper, P. Culicover, M Harnish. (The MIT Press,Cambridge, MA, 1989).
13. M. Buchanan, *Ubiquity.* (Weidenfeld and Nicholson, London, 2000).
14. D.R. Chialvo & P. Bak, Learning from mistakes. *Neuroscience* **90**, 1137(1999).
15. D.R. Chialvo, Critical brain networks, *Physica A* **340**, 756 (2004).
16. D.R. Chialvo, Are our senses critical? *Nature Physics* **2**, 301 (2006).
17. D.R. Chialvo, The brain near the edge. In: *Cooperative behavior in neural systems: Ninth Granada Lectures. AIP Conference Proceedings* **887**,1 (2007).
18. D.R. Chialvo, Emergent complexity: what uphill analysis or downhill invention cannot do. *New Ideas in Psychology* **3**,158 (2008).

19. D.R. Chialvo, Complex emergent neural dynamics *Nature Physics* **6**, 744 (2010).
20. D.R. Chialvo & D. Fraiman, Optimal information-sharing in brain resting state networks. arXiv:1011.1192v1 [q-bio.NC] (2010).
21. D.R. Chialvo & M.M. Millonas, How swarms build cognitive maps. In: *The biology and technology of intelligent autonomous agents.* Luc Steels (ed.) NATO ASI Series, **144**, 439 (1995).
22. J.L. Deneubourg, S. Goss, N. Franks, J.M. Pasteels, The blind leading the blind: Modeling chemically mediated army ant raid patterns. *J. Insect Behavior* **2**, 719 (1989).
23. P. Expert, R. Lambiotte, D.R. Chialvo, K. Christensen, H.J. Jensen, D.J. Sharp, F. Turkheimer, Self-similar correlation function in brain resting state fMRI. arXiv:1003.3682v1 [q-biol.NC] (2010).
24. V.M. Eguiluz, D.R. Chialvo, G. Cecchi, M. Baliki, A.V. Apkarian, Scale free brain functional networks, *Phys. Rev. Lett.* **94**, 018102 (2005).
25. A.A. Fingelkurts & A.A. Fingelkurts, Persistent operational synchrony within brain default-mode network and self-processing operations in healthy subjects. *Brain and Cognition* **75**, 79 (2011).
26. D. Fraiman, P. Balenzuela, J. Foss, D.R. Chialvo, Ising-like dynamics in large scale brain networks. *Phys. Rev. E* **79**, 061922 (2009).
27. M.D. Fox & M.E. Raichle, Spontaneous fluctuations in brain activity observed with functional magnetic resonance imaging. *Nat. Rev. Neurosc.* **8** 700 (2007).
28. M.D. Fox, A.Z. Snyder, J.L. Vincent, M. Corbetta, D.C. van Essen, M.E. Raichle, The human brain is intrinsically organized into dynamic, anticorrelated functional networks. *Proc. Natl. Acad. Sci. U.S.A.* **102**, 9673 (2005).
29. D.D. Garrett, N. Kovacevic, A.R. McIntosh, C.L. Grady, Blood Oxygen Level-Dependent signal variability is more than just noise. *J. Neurosci.***30**, 4914 (2010).
30. D.D. Garrett, N. Kovacevic, A.R. McIntosh, C.L. Grady, The importance of being variable. *J. Neurosci.* in press (2011).
31. O. Kinouchi & M. Copelli, Optimal dynamical range of excitable networks at criticality. *Nature Physics* **2**, 348 (2006).
32. M.G. Kitzbichler, M.L. Smith, S.R. Christensen, E. Bullmore, Broadband criticality of human brain network synchronization. *PLoS. Comput. Biol.* **5**, e1000314(2009).
33. M.M. Millonas, A connectionist type model of self-organized foraging and emergent behavior in ant swarms. *J. Theor. Biol.* **159**, 5299 (1992).
34. M.M. Millonas, Swarms, Phase Transitions, and Collective Intelligence. In: *Artificial Life III.* C.G. Langton (ed.), (Proc. Santa Fe Institute Studies in the Sciences of Complexity, Addison-Wesley, Reading, MA, 1994).
35. M.M. Millonas, A nonequilibrium statistical field theory of swarms and other spatially extended complex systems. In: *Pattern Formation in Physical and Biological Sciences.* P. Claudis, (ed.), (Proc. Santa Fe Institute Studies in the Sciences of Complexity, Addison-Wesley, Reading, MA, 1994).

36. B. Mlslc, T. Mills, M.J. Taylor, A.R. McIntosh, Brain noise is task dependent and region specific. *J. Neurophysiol.* **104**. 2667 (2010).
37. M. Paczuski & P. Bak, Self-organization of complex systems. In: *Proceedings of 12th Chris Engelbrecht Summer School.* Also as cond-mat/9906077.(1999).
38. J.M. Pasteels, J.L. Deneubourg, S. Goss, Transmission and amplification of information in a changing environment: The case of ants. In: *Laws of Nature and Human Conduct.* (Prigogine, I. & Sanglier, M., eds.) (Gordes, Brussels, 1987).
39. T. Petermann, T.C. Thiagarajan, M.A. Lebedev, M.A. Nicolelis, D.R. Chialvo, D. Plenz, Spontaneous cortical activity in awake monkeys composed of neuronal avalanches. *Proc. Natl. Acad. Sci. USA* **106**, 15921 (2009).
40. D. Plenz & D.R. Chialvo, Scaling properties of neuronal avalanches are consistent with critical dynamics. arXiv:0912.5369v1 [q-bio.NC] (2009).
41. D. Plenz & T.C. Thiagarajan, The organizing principles of neuronal avalanche activity: cell assemblies in the cortex? *Trends Neurosci.* textbf3, 101 (2007).
42. E.M. Rauch, M.M. Millonas, D.R. Chialvo, Pattern formation and functionality in swarm models, *Physics Letters A* **207**, 185 (1995).
43. T.L. Ribeiro, M. Copelli, F. Caixeta, H. Belchior, D.R. Chialvo, M.A.L. Nicolelis, S. Ribeiro, Spike avalanches exhibit universal dynamics across the sleep-wake cycle. *PLoS ONE* **5**, e14129 (2010).
44. E. Rolls & G. Deco, *The noisy brain. Stochastic dynamics as a principle of brain function.* (Oxford Univ. Press, UK, 2010)
45. R. Salvador, J. Suckling, M.R. Coleman, J.D. Pickard, D. Menon, E. Bullmore. Neurophysiological architecture of functional magnetic resonance images of human brain. *Cerebral Cortex* **15**, 1332 (2005).
46. S.M. Smith, P.T. Fox, K.L. Miller, D.C. Glahn, P.M. Fox, C.E. Mackay, N. Filippini, K.E. Watkins, R. Toro, A.R. Laird, C.F. Beckmann, Correspondence of the brain's functional architecture during activation and rest. *Proc. Natl. Acad. Sci. USA* **106**, 13040 (2009).
47. R. Sole & B.C. Goodwin, *Signs of Life: How Complexity Pervades Biology* (Basic Books, MA, 2000).
48. O. Sporns, *Networks of the brain* (MIT Press, Cambridge, 2010).
49. O. Sporns, D.R. Chialvo, M. Kaiser, C.C. Hilgetag, Organization, development and function of complex brain networks, *Trends Cog. Sci.* **8**, 418 (2004).
50. C.J. Stam, B.F. Jones, G. Nolte, M. Breakspear, Ph. Scheltens, Small-world networks and functional connectivity in Alzheimer's disease.*Cerebral Cortex* **17**, 92 (2007).
51. S.A. Steyn-Rose & M. Steyn-Rose (eds). *Modeling phase transitions in the brain.* (Springer, NY, 2010).
52. E. Tagliazucchi, P. Balenzuela, D. Fraiman, P. Montoya, D.R. Chialvo, Spontaneous BOLD event triggered averages for estimating functional connectivity at resting state, *Neurosci. Letters* **488**, 158 (2011).
53. G. Tononi, G.M. Edelman, O. Sporns, Complexity and coherency: integrating information in the brain. *Trends Cog. Sci.* **2**, 474 (1998).

54. G. Tononi & G.M. Edelman, Consciousness and complexity. *Science* **282**, 1846 (1998).
55. G. Tononi, An information integration theory of consciousness, *BMC Neurosci.* **5**, 42 (2004).
56. D.L. Turcotte, Self-Organized Criticality, *Reports on Progress in Physics* **62**, 1377 (1999).
57. M.P. van den Heuvel, C.J. Stam, M. Boersma, H.E. Hullshof Pol, Small-world and scale-free organization of voxel-based resting-state functional connectivity in the human brain. *NeuroImage* **43**, 528 (2008).
58. G. Werner, Perspectives on the neuroscience and consciousness. *BioSystems* **87**, 82 (2006).
59. O.E. Wilson, *The Insect Societies.* (Harvard University Press, Cambridge, MA, 1971).

# ACQUIRING LONG-RANGE MEMORY THROUGH ADAPTIVE AVALANCHES

STEFAN BOETTCHER*

*Physics Department, Emory University, Atlanta, GA 30322, USA*

The Bak–Sneppen model demonstrates that extremal selection of poorly adapted variables, which are interconnected in a system, can lead to a highly correlated state that is both, well-adapted and responsive. We have developed a solvable version of the model that provides detailed insights in how memory is acquired in the system. We have employed the Bak–Sneppen dynamics to design a highly efficient optimization heuristic, called "Extremal Optimization" (EO), to find low-energy or low-cost configurations for many combinatorial problems in physics and computer science. Analysis of the local search dynamics induced by EO reveals that the configuration space is traversed most efficiently at the "edge of ergodicity" which effectively exploits the long-range memory encapsulated in the strongly correlated state of the variables. There, the dynamics of local search possesses a generic critical point under the variation of its sole parameter, separating phases of too greedy (non-ergodic, jammed) and too random (ergodic, diffusive) exploration. Analytic comparison with other local search methods, such as a fixed temperature Metropolis algorithm, within this model suggests that the existence of the critical point is the essential distinction leading to the optimal performance of the extremal optimization heuristic.

## 1. Introduction

We have introduced a new heuristic, Extremal Optimization (EO), in Refs. [1,2] and demonstrated its efficiency on a variety of combinatorial [3–6] and physical optimization problems [2,7,8]. Comparative studies with simulated annealing [1,3,9] and other Metropolis based heuristics [10–14] have established EO as a successful alternative for the study of NP-hard problems and its use has spread throughout the sciences. EO has found a large number of applications by other researchers, e.g. for polymer confirmation studies [15,16], pattern recognition [17–19], signal filtering [20,21], transport problems [22], molecular dynamics simulations [23], artificial intelligence [24–26], modeling of social networks [27–29], and $3d$–spin glasses [12,30].

*http://www.physics.emory.edu/faculty/boettcher/

as desired, where the unconditional update ensures significant fluctuations with sufficient incentive to return to near-optimal solutions due to selection *against* variables with poor fitness, for the right choice of $\tau$. Clearly, for finite $\tau$, EO never "freezes" into a single configuration; it is able to return an extensive list of the best of the configurations visited (or simply their cost) instead [5].

For $\tau = 0$, this "$\tau$-EO" algorithm is simply a random walk through configuration space. Conversely, for $\tau \to \infty$, the process approaches a deterministic local search, only updating the lowest-ranked variable, and is likely to reach a dead end. However, for finite values of $\tau$ the choice of a *scale-free* distribution for $P(k)$ in Eq. (4) ensures that no rank gets excluded from further evolution, while maintaining a clear bias against variables with bad fitness. As Sec. 4 will demonstrate, fixing $\tau - 1 \sim 1/\ln(n)$ provides a simple, parameter-free strategy, activating avalanches of adaptation [1,2].

## 4. EO Dynamics

Understanding the Dynamics of EO has proven a useful endeavor [10,47]. Such insights have lead to the implementation of $\tau$-EO described in Sec. 3. Treating $\tau$-EO as an evolutionary process allows us to elucidate its capabilities and to make further refinements. Using simulations, we have analyzed the dynamic pattern of the $\tau$-EO heuristic. As described in Sec. 3, we have implemented $\tau$-EO for the spin glass with Gaussian bonds on a set of instances of 3-regular graphs of sizes $n = 256$, 512, and 1024, and run each instance for $T_{\text{run}} = 20n^3$ update steps. As a function of $\tau$, we measured the ensemble average of the lowest-found energy density $\langle e \rangle = \langle H \rangle / n$, the first-return time distribution $R(\Delta t)$ of update activity to any specific spin, and auto-correlations $C(t)$ between two configurations separated by a time $t$ in a single run. In Fig. 1, we show the plot of $\langle e \rangle$, which confirms the picture found numerically [2,4] and theoretically [47] for $\tau-$EO. The transition at $\tau = 1$ we will investigate further below and theoretically in Sec. 6. The worsening behavior for large $\tau$ has been shown theoretically in Ref. [47] to originate with the fact that in any *finite*-time application, $T_{\text{run}} < \infty$, $\tau-$EO becomes less likely to escape local minima for increasing $\tau$ and $n$. The combination of the purely diffusive search below $\tau = 1$ and the "jammed" state for large $\tau$ leads to the conclusion that the optimal value is approximated by $\tau - 1 \sim 1/\ln(n)$ for $n \to \infty$, consistent with Fig. 1 and experiments in Refs. [2,4].

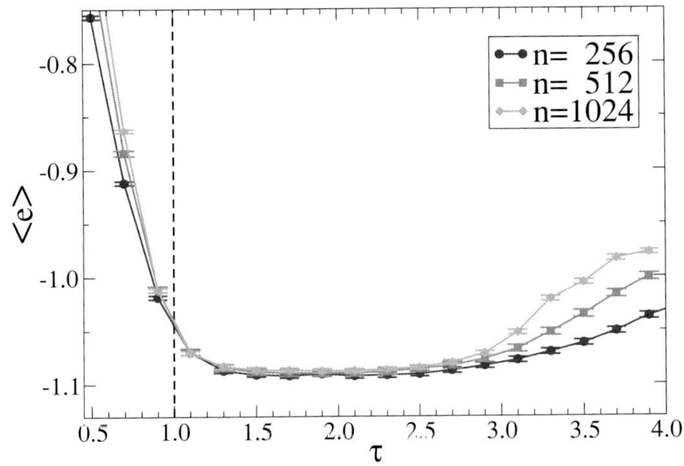


Fig. 1. Plot of the average lowest energy density found with $\tau-$EO over a fixed testbed of 3-regular graph instances of size $n$ for varying $\tau$. For $n \to \infty$, the results are near-optimal only in a narrowing range of $\tau$ just above $\tau = 1$. Below $\tau = 1$ results dramatically worsen, hinting at the phase transition in the search dynamics obtained in Sec. 6.

In Fig. 2 we show the first-return probability for select values of $\tau$. It shows that $\tau$-EO is a fractal renewal process for all $\tau > 1$, and for $\tau < 1$ it is a Poisson process: when variables are drawn according to their "rank" $k$ with probability $P(k)$ in Eq. (3), one gets for the first-return time distribution

$$R(\Delta t) \sim -\frac{P(k)^3}{P'(k)} \sim \Delta t^{\frac{1}{\tau}-2}. \tag{5}$$

Neglecting correlations between variables, the number of updates of a variable of rank $k$ is $\#(k) = T_{\text{run}} P(k)$. Then, the typical life-time is $\Delta t(k) \sim T_{\text{run}}/\#(k) = 1/P(k)$, which via $R(\Delta t) d\Delta t = P(k) dk$ immediately gives Eq. (5). The numerical results in Fig. 2 fit the prediction in Eq. (5) well. Note that the average life-time, and hence the memory preserved by each variable, *diverges* for all $\tau (> 1)$, limited only by $T_{\text{run}}$, a size-dependent cut-off, and is widest for $\tau \to 1^+$, where $\tau$-EO performs optimal. This finding affirms the subtle relation between searching configuration space widely while preserving the memory of good solutions.

Interestingly, the auto-correlations between configurations shown in the inset of Fig. 2 appear to decay with a *stretched*-exponential tail, $C(t) \sim \exp\{-B_\tau \sqrt{t}\}$ fitted with $B_\tau \approx 1.6 \exp\{-2.4\tau\}$, for all $\tau > 1$, characteristic of a super-cooled liquid [46] just *above* the glass transition temperature $T_g (> 0$ in this model). While we have not been able to derive that result, it

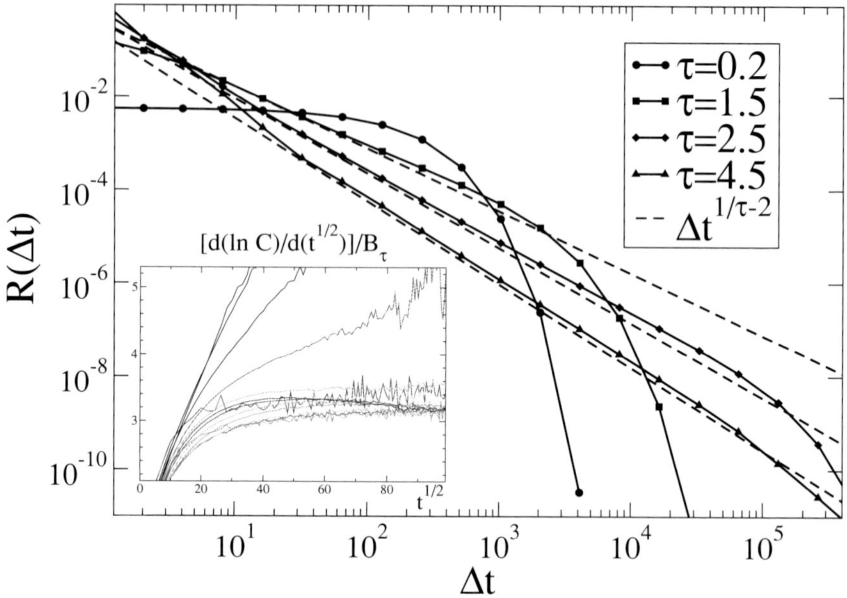


Fig. 2. Plot of the first-return time distribution $R(\Delta t)$ for $\tau$-EO for various $\tau$ and $n = 256$. Poissonian behavior for $\tau < 1$ develops into a power-law regime limited by a cut-off for $\tau > 1$. The power-law scaling closely follows Eq. (5) (dashed lines). Inset: Data collapse (except for $\tau \leq 1$) of autocorrelations $C(t)$ according to the stretched-exponential fit given in the text. From top to bottom, $\tau = 0.5, 0.7, \ldots, 3.5$.

suggests that $\tau$-EO, driven far from equilibrium, never "freezes" into a glassy ($T < T_g$) state, yet accesses $T = 0$ properties efficiently. Such correlations typically decay with an agonizingly anemic power-law [46] for a thermal search of a complex ($T_g > 0$) energy landscape, entailing poor exploration and slow convergence.

## 5. Annealed Optimization Model

As described in Ref. [47], we can abstract certain combinatorial optimization problems into a simple, analytically tractable annealed optimization model (AOM). To motivate AOM, we imagine a generic optimization problem, e.g. the spin glass in Sec. 3, as consisting of a number of variables $1 \leq i \leq n$, each of which contributes an amount $-\lambda_i$ to the overall cost per variable (or energy density),

$$\epsilon = -\frac{1}{n} \sum_{i=1}^{n} \lambda_i. \tag{6}$$

The "fitness" of each variable is $\lambda_i \leq 0$, where larger values are better and $\lambda_i = 0$ is optimal. The (optimal) ground state of the system is $\epsilon = 0$. In a realistic problem, variables are constrained such that not all of them can be simultaneously of optimal fitness. In AOM, those correlations are neglected.

We will consider that each variable $i$ is in one of $\alpha + 1$ different fitness states $\lambda_i$, where $\alpha_i = \alpha$ is fixed as a constant here. (For example, $\alpha = 2d$ on a $d$-dimensional hyper-cubic lattice or $\alpha = 3$ in Sec. 4.) We can specify occupation numbers $n_a$, $0 \leq a \leq \alpha$, for each state $a$, and define occupation densities $\rho_a = n_a/n$ $(a = 0, \ldots, \alpha)$. Hence, any local search procedure with single-variable updates, say, can be cast simply as a set of evolution equations,

$$\dot{\rho}_b - \sum_{a=0}^{\alpha} T_{b,a} Q_a \tag{7}$$

Here, $Q_a$ is the probability that a variable in state $a$ gets updated; any local search process (based on updating a finite number of variables) *defines* a unique set of $\mathbf{Q}$, as we will see below. The matrix $T_{b,a}$ specifies the net transition to state $b$ *given* that a variable in state $a$ is updated. This matrix allows us to *design* arbitrary, albeit annealed, optimization problems for AOM. Both, $\mathbf{T}$ and $\mathbf{Q}$, generally depend on the density vector $\rho(t)$ as well as on $t$ explicitly.

We want to consider the different fitness states equally spaced, as in the spin glass example above, where variables in state $a$ contribute $a\Delta E$ to the energy to the system. Here $\Delta E > 0$ is an arbitrary energy scale. The optimization problem is defined by minimizing the "energy" density

$$\epsilon = \sum_{a=0}^{\alpha} a\rho_a \geq 0. \tag{8}$$

Conservation of probability and of variables implies the constraints

$$\sum_{a=0}^{\alpha} \rho_a(t) = 1, \quad \sum_{a=0}^{\alpha} \dot{\rho}_a = 0, \quad \sum_{a=0}^{\alpha} Q_a = 1, \quad \sum_{a=0}^{\alpha} T_{a,b} = 0 \ (0 \leq b \leq \alpha). \tag{9}$$

While AOM eliminates most of the relevant properties of a truly hard optimization problem, such as quenched randomness and frustration [46], two fundamental features of the evolution equations in Eq. (7) remain appealing: (1) The behavior for a large number of variables can be abstracted into a simple set of equations, describing their dynamics with merely a few unknowns, $\rho$, and (2) the separation of update preference, $\mathbf{Q}$, and update process, $\mathbf{T}$, lends itself to an analytical comparison between different heuristics, as we will show in Sec. 6.

## 6. Evolution Equations for Local Search Heuristics

The AOM developed above is quite generic for a class of combinatorial optimization problems. It was designed in particular to analyze EO [47], which we will review next. Then we will present the update preferences **Q** through which each local search heuristic enters into AOM. We also specify the update preferences **Q** for Metropolis-based local searches, akin to simulated annealing.

### 6.1. *Extremal optimization algorithm*

EO is simply implemented in AOM: For a given configuration $\{x_i\}_{i=1}^n$, assign to each variable $x_i$ a "fitness" $\lambda_i = 0, -1, \ldots, -\alpha$ (e.g. $\lambda_i = -\{\#\textit{violated bonds}\}$ in the spin glass), so that Eq. (6) is satisfied. Each variable falls into one of only $\alpha + 1$ possible states. Say, currently there are $n_\alpha$ variables with the worst fitness, $\lambda = -\alpha$, $n_{\alpha-1}$ with $\lambda = -(\alpha - 1)$, and so on up to $n_0$ variables with the best fitness $\lambda = 0$ with $n = \sum_{b=0}^{\alpha} n_b$. Select an integer $k$, $1 \le k \le n$, from some distribution, preferably with a bias towards lower values of $k$. Determine $0 \le a \le \alpha$ such that $\sum_{b=a+1}^{\alpha} n_b < k \le \sum_{b=a}^{\alpha} n_b$. Note that lower values of $k$ would select a "pool" $n_a$ with larger value of $a$, containing variables of lower fitness. Finally, select one of the $n_a$ variables in state $a$ with equal chance and update it *unconditionally*. As in Eq. (4), we prescribe a bias for selecting variables of poor fitness on a slowly varying (power-law) scale over the *ranking* $1 \le k \le n$ of the variables by their fitnesses $\lambda_i$,

$$P_\tau(k) = \frac{\tau - 1}{1 - n^{1-\tau}}\, k^{-\tau} \quad (1 \le k \le n). \tag{10}$$

As an alternative, we can also study EO with threshold updating, which Ref. [35] has shown rigorously to be optimal. Yet, the actual value of this threshold at any point in time is typically not obvious (see also Ref. [36]). We will implement a sharp threshold $s$ ($1 \le s \le n$) via

$$P_s(k) \propto \frac{1}{1 + e^{r(k-s)}} \quad (1 \le k \le n) \tag{11}$$

for $r \to \infty$. Since we can only consider fixed thresholds $s$, it is not apparent how to shape the rigorous results into a successful algorithm.

### 6.2. *Update probabilities for extremal optimization*

As described in Sec. 6.1, in each update of $\tau$-EO a variable is selected based on its rank according to the probability distribution in Eq. (10). When a rank $k(\leq n)$ has been chosen, a variable is randomly picked from state $\alpha$, if $k/n \leq \rho_\alpha$, from state $\alpha - 1$, if $\rho_\alpha < k/n \leq \rho_\alpha + \rho_{\alpha-1}$, and so on. We introduce a new, continuous variable $x = k/n$, for large $n$ approximate sums by integrals, and rewrite $P(k)$ in Eq. (10) as

$$p_\tau(x) = \frac{\tau - 1}{n^{\tau-1} - 1} x^{-\tau} \quad \left(\frac{1}{n} \leq x \leq 1\right), \tag{12}$$

where the maintenance of the low-$x$ cut-off at $1/n$ will turn out to be crucial. Now, the average likelihood in EO that a variable in a given state is updated is given by

$$\begin{aligned}
Q_\alpha &= \int_{1/n}^{\rho_\alpha} p(x)dx = \frac{1}{1 - n^{\tau-1}} \left(\rho_\alpha^{1-\tau} - n^{\tau-1}\right), \\
Q_{\alpha-1} &= \int_{\rho_\alpha}^{\rho_\alpha+\rho_{\alpha-1}} p(x)dx = \frac{1}{1 - n^{\tau-1}} \left[\left(\rho_{\alpha-1} + \rho_\alpha\right)^{1-\tau} - \rho_\alpha^{1-\tau}\right], \\
&\dots \\
Q_0 &= \int_{1-\rho_0}^{1} p(x)dx = \frac{1}{1 - n^{\tau-1}} \left[1 - \left(1 - \rho_0\right)^{1-\tau}\right],
\end{aligned} \tag{13}$$

where in the last line the norm $\sum_a \rho_a = 1$ was used. These values of the $Q$'s completely describe the update preferences for $\tau$-EO at arbitrary $\tau$.

Similarly, we can proceed with the threshold distribution in Eq. (11) to obtain

$$p_s(x) \propto \frac{1}{1 + e^{r(nx-s)}} \quad (\frac{1}{n} \leq x \leq 1), \tag{14}$$

with some proper normalization. While all the integrals to obtain $\mathbf{Q}$ in Eq. (13) are elementary, we do not display the rather lengthy results here.

Note that all the update probabilities in each variant of EO are *independent* of the matrix $\mathbf{T}$ in Eq. (7), i.e. of any particular model, which remains to be specified. This is special, as the following case of a Metropolis algorithm shows.

### 6.3. *Update probabilities for metropolis algorithms*

It is more difficult to construct $\mathbf{Q}$ for a Metropolis-based algorithm, like simulated annealing [48,49]. Let's assume that we consider a variable in state $a$ for an update. Certainly, $Q_a$ would be proportional to $\rho_a$, since variables are randomly selected for an update. But as the actual update of the chosen variable may be accepted or rejected based on a Metropolis condition, further considerations are necessary. The requisite Boltzmann factor $e^{-\beta n \Delta \epsilon_a}$ for the potential update from time $t \to t+1$ of a variable in $a$, aside from the inverse temperature $\beta(t)$, only depends on the entries for $T_{a,b}$:

$$\Delta\epsilon_a = \sum_{b=0}^{\alpha} b\left[\rho_b(t+1)-\rho_b(t)\right]\Big|_a \sim \sum_{b=0}^{\alpha} b\dot{\rho}_b\Big|_a = \sum_{b=0}^{\alpha} b\sum_{c=0}^{\alpha} T_{b,c}Q_c\Big|_a = \sum_{b=0}^{\alpha} bT_{b,a},$$

where the subscript $a$ expresses the fact that it is a *given* variable in state $a$ considered for an update, i.e. $Q_c|_a = \delta_{a,c}$. Hence, from Metropolis we find for the average probability of an update of a variable in state $a$:

$$Q_a = \frac{1}{\mathcal{N}}\,\rho_a \min\left\{1, \exp\left[-\beta n \sum_{b=0}^{\alpha} bT_{b,a}\right]\right\}, \tag{15}$$

where the norm $\mathcal{N}$ is determined via $\sum_a Q_a = 1$. Unlike for EO, the update probabilities here are model-specific, i.e. they depend on the matrix $\mathbf{T}$.

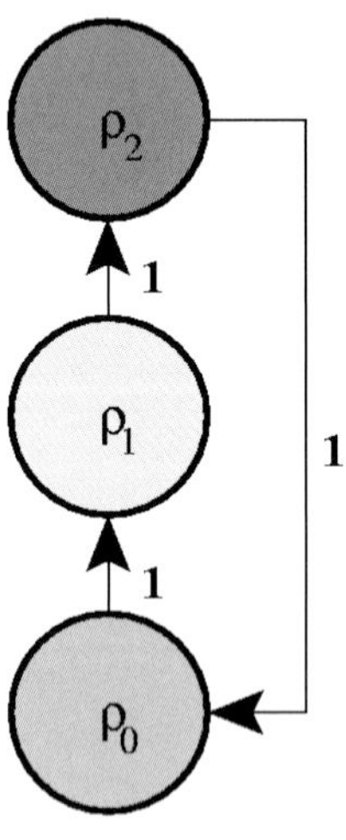


**Flow up**

Fig. 3. Flow diagram with energetic barriers. Arrows indicate the net number of variables transferred, $nT_{b,a}$, into a state $b$, given that a variable in $a$ gets updated. Diagonal elements $T_{a,a}$ correspondingly are negative, accounting for the outflow. Here, variables transferring from $\rho_1$ to $\rho_0$ must first jump up in energy to $\rho_2$.

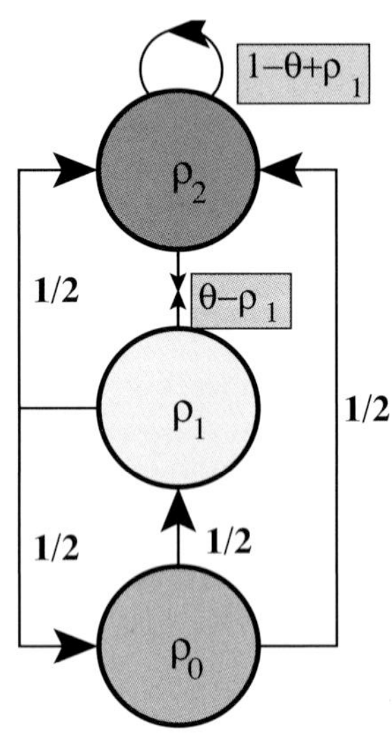


**Flow jam**

Fig. 4. Same as Fig. 3, but with a model leading to a jam. Variables can only transfer from $\rho_2$ to $\rho_0$ through $\rho_1$, but only if $\rho_1 < \theta$. Once $\rho_1 = \theta$, flow down from $\rho_2$ ceases until $\rho_1$ reduces again.

### 6.4. *Evolution equations for a simple barrier model*

To demonstrate the use of these equations, we consider a simple model of an energetic barrier with only three states ($\alpha = 2$) and a constant flow matrix $T_{b,a} = [-\delta_{b,a} + \delta_{(2+b \bmod 3),a}]/n$, depicted in Fig. 3. Here, variables in $\rho_1$ can only reach their lowest-energy state in $\rho_0$ by first jumping *up* in energy to $\rho_2$. Eq. (7) gives

$$\dot{\rho}_0 = \frac{1}{n}\left(-Q_0 + Q_2\right), \quad \dot{\rho}_1 = \frac{1}{n}\left(Q_0 - Q_1\right), \quad \dot{\rho}_2 = \frac{1}{n}\left(Q_1 - Q_2\right), \tag{16}$$

with some $\mathbf{Q}$ discussed in Sec. 6.2 for the variants of EO.

Given this $\mathbf{T}$, we can now also determine the specific update probabilities for Metropolis according to Eqs. (15). Note that for $a = 2$ we can evaluate the *min* as 1, since $\sum_b bT_{b,a=2} < 0$ always, while for $a = 0, 1$ the *min* always evaluates to the exponential. Properly normalized, we obtain

$$Q_0 = \frac{\rho_0 e^{-\beta/2}}{(1-e^{-\beta/2})\rho_2 + e^{-\beta/2}}, \quad Q_1 = \frac{\rho_1 e^{-\beta/2}}{(1-e^{-\beta/2})\rho_2 + e^{-\beta/2}},$$
$$Q_2 = \frac{\rho_2}{(1-e^{-\beta/2})\rho_2 + e^{-\beta/2}}. \tag{17}$$

It is now very simple to obtain the stationary solution: For $\dot{\rho} = 0$, Eqs. (16) yield $Q_0 = Q_1 = Q_2 = 1/3$, and we obtain from Eq. (13) for $\tau$-EO:

$$\rho_0 = 1 - \left(\frac{1}{3}n^{\tau-1} + \frac{2}{3}\right)^{\frac{1}{1-\tau}}, \quad \rho_2 = \left(\frac{2}{3}n^{\tau-1} + \frac{1}{3}\right)^{\frac{1}{1-\tau}}, \tag{18}$$

and $\rho_1 = 1 - \rho_0 - \rho_2$, and for Metropolis:

$$\rho_0 = \frac{1}{2+e^{-\beta/2}}, \quad \rho_1 = \frac{1}{2+e^{-\beta/2}}, \quad \rho_2 = \frac{e^{-\beta/2}}{2+e^{-\beta/2}}. \tag{19}$$

For EO with threshold updating, we obtain

$$\begin{aligned}
\rho_0 &= \frac{1}{3} - \frac{1}{3n} - \frac{s}{n} - \frac{1}{3nr}\ln\left[1 + e^{r(n-s)}\right] \\
&\quad + \frac{1}{nr}\ln\left[\left(e^{nr} + e^{rs}\right)\left(1 + e^{r(1-s)}\right)^{\frac{1}{3}} + e^{\frac{r}{3}(2n+1)}\left(1 + e^{r(n-s)}\right)^{\frac{1}{3}}\right], \\
\rho_2 &= \frac{1}{3} + \frac{2}{3n} + \frac{s}{n} - \frac{2}{3nr}\ln\left[1 + e^{r(n-s)}\right] \\
&\quad + \frac{1}{nr}\ln\left[\left(e^{nr} + e^{rs}\right)\left(1 + e^{r(1-s)}\right)^{\frac{2}{3}} + e^{\frac{r}{3}(n+2)}\left(1 + e^{r(n-s)}\right)^{\frac{2}{3}}\right],
\end{aligned} \tag{20}$$

and, assuming a threshold anywhere between $1 < s < n$, for $r \to \infty$:

$$\rho_0 = 1 - \frac{2s+1}{3n}, \quad \rho_2 = \frac{s+2}{3n}, \quad \rho_1 = \frac{s-1}{3n}. \tag{21}$$

Therefore, according to Eq. (8), Metropolis reaches its best, albeit sub-optimal, cost $\epsilon = 1/2 > 0$ at $\beta \to \infty$, due to the energetic barrier faced by the variables in $\rho_1$, see Fig. 3. The result for threshold updating in EO are more promising: near-optimal results are obtained, to within $O(1/n)$, for any finite threshold $s$. But results are best for small $s \to 1 \ll n$, in which limit we revert back to "basic" EO (only update the worst) obtained also for $\tau \to \infty$.

Finally, the result for $\tau$-EO is most remarkable: For $n \to \infty$ at $\tau < 1$ EO remains sub-optimal, but reaches the optimal cost in the *entire* domain $\tau > 1$! This transition at $\tau = 1$ separates a (diffusive) random walk phase with too much fluctuation, and a greedy descent phase with too little fluctuation, which would trap $\tau$-EO in problems with a complex landscape. This transition derives *generically* from the scale-free power-law in Eq. (10), as had been argued on the basis of numerical results for real NP-hard problems in Refs. [2,4]. The difference between reaching optimality in a limit only ($\tau \to \infty$ as in basic EO, $s \to 1$ for our naive threshold-EO model) or within a phase ($\tau > 1$) seems insignificant in the stationary regime, $T_{\rm run} \to \infty$. Yet, it is the hallmark of a local search in a complex landscape that stationarity is rarely reached within any reasonable computational time $T_{\rm run} < \infty$. At *intermediate* times, constrained variables jam each others evolution, requiring a subtle interplay between greedy descent and activated fluctuations to escape metastable states, as we will analyze in the following.

### 6.5. *Jamming model for $\tau$-EO*

Naturally, the range of phenomena found in a local search of NP-hard problems is not limited to energetic barriers. After all, so far we have only considered constant entries for $T_{b,a}$. Therefore, as our next AOM we want to review the case of $\mathbf{T}$ depending linearly on the $\rho_i$ for $\tau$-EO [47]. It highlights the importance of the fact that $\tau$-EO attains optimality in the entire phase $\tau > 1$, instead of just an extreme limit such as basic EO ($\tau \to \infty$) or $s \to 1$ for EO with fixed threshold. From Fig. 4, we can read off $\mathbf{T}$ and obtain for Eq. (7):

$$\dot{\rho}_0 = \frac{1}{n}\left[-Q_0 + \frac{1}{2}Q_1\right],$$
$$\dot{\rho}_1 = \frac{1}{n}\left[\frac{1}{2}Q_0 - Q_1 + (\theta - \rho_1)Q_2\right], \tag{22}$$

and $\dot{\rho}_2 = -\dot{\rho}_0 - \dot{\rho}_1$ from Eq. (9). Aside from the dependence of $\mathbf{T}$ on $\rho_1$, we have also introduced the threshold parameter $\theta$. In fact, if $\theta \geq 1$, the model behaves effectively like the previous model, and for $\theta \leq 0$ there can be no flow from state 2 to the lower states at all. The interesting regime is the case $0 < \theta < 1$, where further flow from state 2 into state 1 can be blocked for increasing $\rho_1$, providing a negative feed-back to the system. In effect, the model is capable of exhibiting a "jam" as observed in many models of glassy dynamics [46], and which is certainly an aspect of local search processes. Indeed, the emergence of such a jam is characteristic of the low-temperature properties of spin glasses and real optimization problems: After many update steps, most variables freeze into a near-perfect local arrangement and resist further change, while a finite fraction remains frustrated (temporarily in this model, permanently in real problems) in a poor local arrangement [50]. More and more of the frozen variables have to be dislodged collectively to accommodate the frustrated variables before the system as a whole can improve its state. In this highly correlated state, frozen variables block the progression of frustrated variables, and a jam emerges.

We obtain for the steady state, $\dot{\rho} = 0$:

$$0 = \frac{3}{2}(A-1) + \left[\theta - A^{1/(1-\tau)} + (3A-2)^{1/(1-\tau)}\right]\left(3A - 2 - n^{\tau-1}\right), \tag{23}$$

which can be solved for $A$ to obtain

$$\rho_0 = 1 - A^{1/(1-\tau)}, \qquad \rho_2 = (3A-2)^{1/(1-\tau)}, \tag{24}$$

and $\rho_1 = 1 - \rho_0 - \rho_2$. Eq. (23) has a unique physical solution ($A > 2/3$) for the $\rho$'s for all $0 \leq \tau \leq \infty$, $0 < \theta < 1$, and all $n$. As advertised in Sec. 6.4, in the thermodynamic limit $n \to \infty$ the generic critical point of $\tau$–EO at $\tau = 1$ emerges: If $\tau < 1$, the sole $n$-dependent term in Eq. (23) vanishes, allowing $A$, and hence the $\rho$'s, to take on finite values, i.e. $e > 0$ in Eq. (8). If $\tau > 1$, the $n$-dependent term diverges, forcing $A$ to diverge in kind, resulting in $\rho_0 \to 1$ and $\rho_i \to 0$ for $i > 0$ in Eqs. (24), i.e. $e \to 0$.

While the steady state ($t \to \infty$) features of this model do not seem to be much different from the model in Sec. 6.4, the dynamics at intermediate times $t$ is more subtle. In particular, as was shown in Ref. [47], a jam in the flow of variables towards better fitness may ensue under certain circumstances. The emergence of the jam depends on initial conditions, and its duration will prove to get longer for larger values of $\tau$. If the initial conditions place a fraction $\rho_0 > 1 - \theta$ already into the lowest state, most likely no jam will emerge, since $\rho_1(t) < \theta$ for all times, and the ground state is reached in $t = O(n)$ steps. But if initially $\rho_1 + \rho_2 = 1 - \rho_0 > \theta$, and $\tau$ is sufficiently large, $\tau$-EO will drive the system to a situation where $\rho_1 \approx \theta$ by preferentially transferring variables from $\rho_2$ to $\rho_1$, as Fig. 5 shows. Then, further evolution becomes extremely slow, delayed by the $\tau$-dependent, small probability that a variable in state 1 is updated ahead of an *extensive* ($\propto n$) number of less-fit variables in state 2.

Clearly, this jam is *not* a stationary solution of Eq. (22). We consider initial conditions leading to a jam, $\rho_1(0) + \rho_2(0) > \theta$ and make the Ansatz

$$\rho_1(t) = \theta - \eta(t) \tag{25}$$

with $\eta \ll 1$ for $t \lesssim t_{\rm jam}$, where $t_{\rm jam}$ is the time before $\rho_0 \to 1$. To determine $t_{\rm jam}$, we apply Eq. (25) to the evolution equations in (22) and use the norm

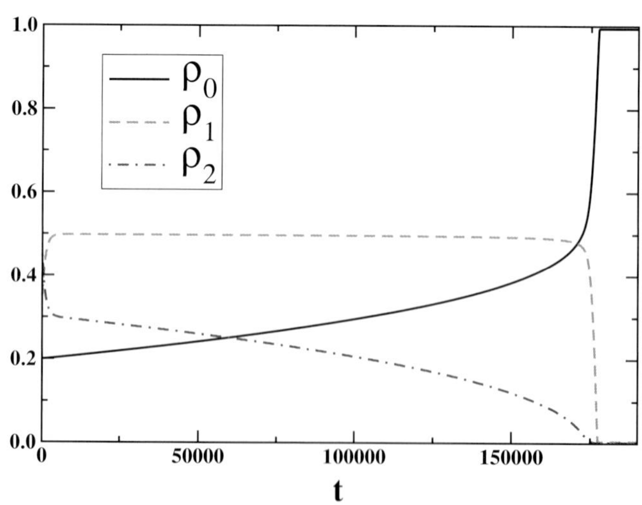


Fig. 5. Plot of the typical evolution of the system in Eqs. (22) for some generic initial condition that leads to a jam. Shown are $\rho_0(t)$, $\rho_1(t)$, and $\rho_2(t)$ for $n = 1000$, $\tau = 2$, $\theta = 0.5$ and initial conditions $\rho_0(0) = 0.2$, $\rho_1(0) = \rho_2(0) = 0.4$. Since $\rho_1(0) < \theta$, $\rho_1$ fills up to $\theta$ almost instantly with variables from $\rho_2$ while $\rho_0$ stays $\approx$constant. After that, $\rho_1 \approx \theta$ for a very long time ($\gg n$) while variables slowly trickle down through state 1. Eventually, after $t = O(n^\tau)$, $\rho_2$ vanishes and EO can empty out $\rho_1$ directly which leads to the ground state $\rho_0 = 1$ (i.e. $e = 0$) almost instantly.

and $\dot{\rho}_1 = 0$ to leading-order, $\dot{\rho}_0 = -\dot{\rho}_2$, which yields an equation solely for $\rho_2(t)$,

$$-\frac{d\rho_2}{dt} \sim \frac{1}{n^\tau}\left[1 - \frac{3}{2}(\theta + \rho_2)^{1-\tau} + \frac{1}{2}\rho_2^{1-\tau}\right], \tag{26}$$

or, using the fact that $\rho_2$ almost instantly takes on the value of $\rho_1(0) + \rho_2(0) - \theta = 1 - \theta - \rho_0(0)$ (see Fig. 5), we solve Eq. (26) to get

$$t \sim n^\tau \int_{\rho_2(t)}^{1-\theta-\rho_0(0)} \frac{2d\xi}{2 - 3(\theta+\xi)^{1-\tau} + \xi^{1-\tau}}. \tag{27}$$

We can estimate the duration of the jam ending at $t = t_{\rm jam}$ by setting $\rho_2(t_{\rm jam}) \approx 0$, see Fig. 5, leaving the integral as a constant to find:

$$t_{\rm jam} \sim n^\tau. \tag{28}$$

Instead of repeating the lengthy calculation in Ref. [47] for the ground state energy averaged over all possible initial conditions for finite runtime $T_{\rm run} \propto n$, we can content ourselves here with the obvious remark that a finite fraction of the initial conditions will lead to a jam, hence will require a runtime $T_{\rm run} \gg t_{\rm jam}$ to reach optimality. With $T_{\rm run} \propto n^k$, the fact that the phase transition in $\tau$-EO provides good solutions for all $\tau > 1$ allows

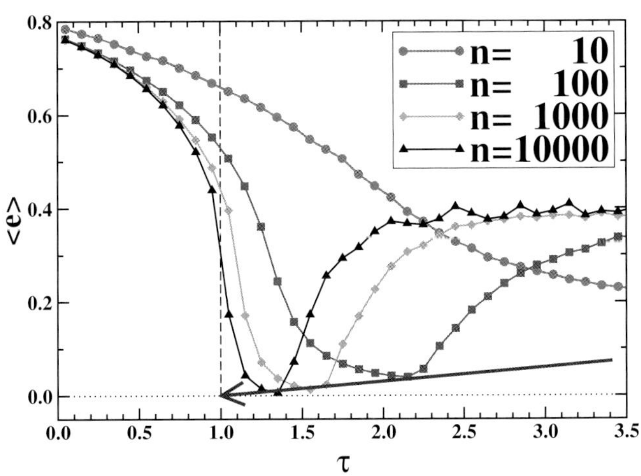


Fig. 6. Plot of the energy $\langle e \rangle$, averaged over initial conditions, vs. $\tau$ in many $\tau$-EO runs of Eqs. (22) with $T_{\rm run} = 100n$, $n = 10$, 100, 1000, and 10000 and $\theta = 1/2$. For small values of $\tau$, $\langle e \rangle$ closely follows the steady state solutions from Eqs. (23-24). It reaches a minimum at a value near the prediction for $\tau_{\rm opt} \approx 3.5$, 2.1, 1.6, and 1.4, approaching $\tau = 1^+$ along the long arrow, and rises sharply beyond that, comparable to Fig. 1.

us to choose $1 < \tau < k$, as is apparent from Fig. 1 where $k = 3$. Fig. 6, obtained here from simulations of this jammed model in Eqs. (22), verifies the general asymptotic scaling, $\tau_{\rm opt} - 1 \sim 1/\ln(n)$, with small enough $\tau$ to fluctuate out of any jam in a time near-linear in $n$ while still attaining optimal results as it would for *any* $\tau > 1$ at infinite runtime.

## References

[1] S. Boettcher and A.G. Percus, "Nature's way of optimizing," *Artificial Intelligence*, vol. 119, p. 275, 2000.
[2] S. Boettcher and A.G. Percus, "Optimization with extremal dynamics," *Phys. Rev. Lett.*, vol. 86, pp. 5211–5214, 2001.
[3] S. Boettcher and A.G. Percus, "Extremal optimization: Methods derived from co-evolution," in *GECCO-99: Proceedings of the Genetic and Evolutionary Computation Conference*, (Morgan Kaufmann, San Francisco), pp. 825–832, 1999.
[4] S. Boettcher and A.G. Percus, "Extremal optimization for graph partitioning," *Phys. Rev. E*, vol. 64, p. 026114, 2001.
[5] S. Boettcher and A.G. Percus, "Extremal optimization at the phase transition of the 3-coloring problem," *Phys. Rev. E*, vol. 69, p. 066703, 2004.
[6] H.H. Hoos and T. Stützle, *Stochastic Local Search: Foundations and Applications.* San Francisco: Morgan Kaufmann, 2004.
[7] S. Boettcher, "Numerical results for ground states of mean-field spin glasses at low connectivities," *Phys. Rev. B*, vol. 67, p. R060403, 2003.
[8] S. Boettcher, "Extremal optimization for Sherrington-Kirkpatrick spin glasses," *Eur. Phys. J. B*, vol. 46, pp. 501–505, 2005.
[9] S. Boettcher, "Extremal optimization and graph partitioning at the percolation threshold," *J. Math. Phys. A: Math. Gen.*, vol. 32, pp. 5201–5211, 1999.
[10] S. Boettcher and M. Frank, "Optimizing at the ergodic edge," *Physica A*, vol. 367, pp. 220–230, 2006.
[11] S. Boettcher and P. Sibani, "Comparing extremal and thermal explorations of energy landscapes," *Eur. Phys. J. B*, vol. 44, pp. 317–326, 2005.
[12] J. Dall and P. Sibani, "Faster monte carlo simulations at low temperatures: The waiting time method," *Computer Physics Communication*, vol. 141, pp. 260–267, 2001.
[13] J. Wang, "Transition matrix Monte Carlo and flat-histogram algorithm," in *AIP Conf. Proc. 690: The Monte Carlo Method in the Physical Sciences*, pp. 344–348, 2003.
[14] J.-S. Wang and Y. Okabe, "Comparison of Extremal Optimization with flat-histogram dynamics for finding spin-glass ground states," *J. Phys. Soc. Jpn.*, vol. 72, p. 1380, 2003.
[15] N.G. Mang and C. Zeng, "Reference energy extremal optimization: A stochastic search algorithm applied to computational protein design," *J. Comp. Chem.*, vol. 29, pp. 1762–1771, 2008.

[16] A. Shmygelska, "An extremal optimization search method for the protein folding problem: the go-model example," in *GECCO '07: Proceedings of the 2007 GECCO conference companion on Genetic and evolutionary computation*, (New York, NY, USA), pp. 2572–2579, ACM, 2007.

[17] S. Meshoul and M. Batouche, "Ant colony system with extremal dynamics for point matching and pose estimation," in *16th International Conference on Pattern Recognition (ICPR'02)*, vol. 3, p. 30823, 2002.

[18] S. Meshoul and M. Batouche, "Robust point correspondence for image registration using optimization with extremal dynamics," *Lecture Notes in Computer Science*, vol. 2449, pp. 330–337, 2002.

[19] S. Meshoul and M. Batouche, "Combining Extremal Optimization with singular value decomposition for effective point matching," *Int. J. Pattern Rec. and AI*, vol. 17, pp. 1111–1126, 2003.

[20] P. Svenson, "Extremal Optimization for sensor report pre-processing," *Proc. SPIE*, vol. 5429, pp. 162–171, 2004.

[21] E. Yom-Tov, A. Grossman, and G.F. Inbar, "Movement-related potentials during the performance of a motor task i: The effect of learning and force," *Biological Cybernatics*, vol. 85, pp. 395–399, 2001.

[22] F.L. de Sousa, V. Vlassov, and F.M. Ramos, "Heat pipe design through generalized extremal optimization," *Heat Transf. Eng.*, vol. 25, pp. 34–45, 2004.

[23] T. Zhou, W.-J. Bai, L.-J. Cheng, and B.-H. Wang, "Continuous Extremal Optimization for Lennard-Jones clusters," *Phys. Rev. E*, vol. 72, p. 016702, 2005.

[24] M.E. Menai and M. Batouche, "Extremal Optimization for Max-SAT.," in *Proceedings of the International Conference on Artificial Intelligence (IC-AI)*, pp. 954–958, 2002.

[25] M.E. Menai and M. Batouche, "A Bose-Einstein Extremal Optimization method for solving real-world instances of maximum satisfiablility.," in *Proceedings of the International Conference on Artificial Intelligence (IC-AI)*, pp. 257–262, 2003.

[26] M.E. Menai and M. Batouche, "Efficient initial solution to Extremal Optimization algorithm for weighted MAXSAT problem," *Lecture Notes in Computer Science*, vol. 2718, pp. 592–603, 2003.

[27] L. Danon, A. Diaz-Guilera, J. Duch, and A. Arenas, "Comparing community structure identification," *J. Stat. Mech.-Theo. Exp.*, p. P09008, 2005.

[28] J. Duch and A. Arenas, "Community detection in complex networks using Extremal Optimization," *Phys. Rev. E*, vol. 72, p. 027104, 2005.

[29] Z. Neda, R. Florian, M. Ravasz, A. Libal, and G. Györgyi, "Phase transition in an optimal clusterization model," *Physica A.*, vol. 362, pp. 357–368, 2006.

[30] R.N. Onody and P.A. de Castro, "Optimization and self-organized criticality in a magnetic system," *Physica A*, vol. 322, pp. 247–255, 2003.

[31] F.L. de Sousa, F.M. Ramos, R.L. Galski, and I. Muraoka, "Generalized extremal optimization: A new meta-heuristic inspired by a model of natural evolution," *Recent Developments in Biologically Inspired Computing*, 2004.

[32] F.L. de Sousa, V. Vlassov, and F.M. Ramos, "Generalized Extremal Optimization for solving complex optimal design problems," *Lecture Notes in Computer Science*, vol. 2723, pp. 375–376, 2003.
[33] M. Iwamatsu and Y. Okabe, "Basin hopping with occasional jumping," *Chem. Phys. Lett.*, vol. 399, pp. 396–400, 2004. cond-mat/0410723.
[34] A.A. Middleton, "Improved Extremal Optimization for the Ising spin glass," *Phys. Rev. E*, vol. 69, p. 055701(R), 2004.
[35] F. Heilmann, K.H. Hoffmann, and P. Salamon, "Best possible probability distribution over Extremal Optimization ranks," *Europhys. Lett.*, vol. 66, pp. 305–310, 2004.
[36] K.H. Hoffmann, F. Heilmann, and P. Salamon, "Fitness threshold accepting over Extremal Optimization ranks," *Phys. Rev. E*, vol. 70, p. 046704, 2004.
[37] A.K. Hartmann and H. Rieger, eds., *New Optimization Algorithms in Physics*. Berlin: Wiley-VCH, 2004.
[38] S.A. Kauffman and S. Johnsen, "Coevolution to the edge of chaos: Coupled fitness landscapes, poised states, and coevolutionary avalanches," *J. Theor. Biol.*, vol. 149, pp. 467–505, 1991.
[39] P. Bak, *How Nature Works* (Springer, New York, 1996).
[40] P. Bak, C. Tang, and K. Wiesenfeld, *Self-Organized Criticality,* Phys. Rev. Lett. **59**, 381 (1987).
[41] P. Bak and K. Sneppen, *Punctuated Equilibrium and Criticality in a simple Model of Evolution,* Phys. Rev. Lett. **71**, 4083-4086 (1993).
[42] D.M. Raup and J.J. Sepkoski, *Periodic Extinction of Families and Genera,* Science **231**, 833-836 (1986).
[43] S.J. Gould and N. Eldridge, *Punctuated Equilibria: The Tempo and Mode of Evolution Reconsidered,* Paleobiology **3**, 115-151 (1977).
[44] S. Boettcher and M. Paczuski, *Ultrametricity and Memory in a Solvable Model of Self-Organized Criticality,* Physical Review E **54**, 1082 (1996).
[45] A. Percus, G. Istrate, and C. Moore, *Computational Complexity and Statistical Physics*. New York: Oxford University Press, 2006.
[46] K.H. Fischer and J.A. Hertz, *Spin Glasses*. Cambridge: Cambridge University Press, 1991.
[47] S. Boettcher and M. Grigni, "Jamming model for the extremal optimization heuristic," *J. Phys. A: Math. Gen.*, vol. 35, pp. 1109–1123, 2002.
[48] S. Kirkpatrick, C.D. Gelatt, and M.P. Vecchi, "Optimization by simulated annealing," *Science*, vol. 220, pp. 671–680, 1983.
[49] P. Salamon, P. Sibani, and R. Frost, *Facts, Conjectures, and Improvements for Simulated Annealing*. Society for Industrial & Applied Mathematics, 2002.
[50] R.G. Palmer, D.L. Stein, E. Abraham, and P.W. Anderson, "Models of hierarchically constrained dynamics for glassy relaxation," *Phys. Rev. Lett.*, vol. 53, pp. 958–961, 1984.

# RANDOM WALK ON COMPLEX NETWORKS: FROM INFINITELY SLOW TO THE INSTANTANEOUS TRANSITION TO EQUILIBRIUM*

NICHOLAS W. HOLLINGSHAD
and
PAOLO GRIGOLINI
*Center for Nonlinear Science, University of North Texas, P.O. Box 311427, Denton, Texas 76203-1427 USA*

PAOLO ALLEGRINI
*Instituto di Fisiologia Clinica (IFC-CNR), Via Moruzzi 1 56124 Pisa, Italy*
*Centro EXTREME, Scuola Superiore Sant'Anna, Piazza Martiri della Libertà 7, 56127 Pisa, Italy*

We study random walks on networks, either regular or complex, and we show that the observation of a single walker is not quite adequate to properly determine the network efficiency, which is expected to increase with the network topological complexity. In fact, the time complexity of a single walker is compatible with the two opposite conditions, of regular topology, with minimum efficiency, and of scale-free topology, with rich nodes that are linked to all the other nodes of the network, thereby generating the condition of maximum efficiency. The observation of the motion of a large number of walkers leads to unambiguous results, with the former limiting condition yielding equilibrium with a time scale proportional to the number of nodes, $N$, and the latter limiting condition corresponding to an almost instantaneous transition to equilibrium, more precisely with a time scale proportional to $\log N$.

## 1. Introduction

Complex networks are a subject of increasing interest that has been illustrated in many recent books [1]. A big emphasis has been given to the discovery that real networks depart from the randomness condition

---

*This work is supported by ARO Grant No. W911NU-05-1-0205.

assumed in the classical work of Erdös and Rényi in 1959, 1960 and 1961. For an accurate reference to the work of these two outstanding mathematicians we refer the readers to the review paper [2]. Here we limit ourselves to point out that in the recent literature there is a widespread conviction that the departure from the random condition is determined by a natural evolution towards the condition of maximal efficiency. Efficiency is evaluated in terms of the minimum numbers of links connecting a given node $i$ to another node $j$ [3]. The definition of efficiency [3] is based on the intuitive notion that the transport of information requires a walker to move from the node $i$ to deliver a message to a node $j$. A network fitting the condition of fast message delivery was denoted as *small world* by Watts and Strogatz [4], and the idea of evaluating the network efficiency through message delivery traces back to the pioneer work of Milgram [5] (see also [6]).

In his popular experiment, Milgram sent several packages to 160 random people living in Omaha, Nebraska, asking them to forward the package to a friend or acquaintance whom they thought would bring the package closer to a set final individual, a stockbroker from Boston, Massachusetts.

The letter included this specific condition: "If you do not know the target person on a personal basis, do not try to contact him directly. Instead, mail this folder to a personal acquaintance who is more likely than you to know the target person." The outcome of the experiment revealed that, without any global network knowledge, letters reached the target recipient using, on average, 5.2 intermediate people, demonstrating that social acquaintance networks were indeed small worlds, of size $L \approx 5$.

On the basis of these remarks the random work becomes an important tool to discuss the network efficiency. This is the reason why authors [7,8] proposed the adoption of the random walk method to discuss the properties of complex networks. The main purpose of the present paper is to study again the random walk approach. By revisiting this problem, we reach the conclusion that the random walk procedure advocated by certain authors [7,8] is an adequate method to assess the network efficiency provided that the Gibbs method is used: We have to study the time evolution of a large number of walkers rather than the time properties of a single random walker. The adoption of the Gibbs perspective leads us to define the equilibrium condition, namely, the constant in time fraction of walkers located on each of the $N$ nodes of the network. We determine that the adoption of this prescription leads to the result: The transition to equilibrium becomes faster with increasing network efficiency, as measured by Ref. [3].

Note that to clarify our arguments we study networks of very small size, $L = 2$. Thus, we find a regression to equilibrium even faster than that assessed experimentally by Milgram.

## 2. Preliminary Remarks on the Size of a Complex Network

In this paper we focus attention mainly on three kinds of networks. The first is the popular model proposed by Watts and Strogatz [4] illustrated by Fig. 1 and referred to herein as the WS model. We see that the condition $p = 0$ generates a condition where the attainment of equilibrium is very slow insofar as the time necessary to get equilibrium is proportional to $N$, the number of nodes of the network. The increase of $p$ generates long-range links that have the effect of reducing the network

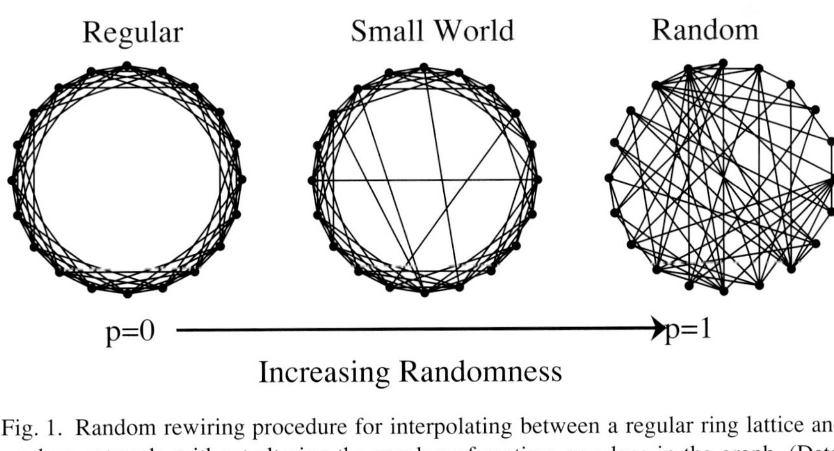


Fig. 1. Random rewiring procedure for interpolating between a regular ring lattice and a random network, without altering the number of vertices or edges in the graph. (Details are provided in [4].) Three realizations of this process are shown, for different values of $p$. For $p = 0$, the network is regular; as $p$ increases, the graph becomes increasingly disordered until for $p = 1$, all edges are rewired randomly. The main result is that for intermediate values of $p$, the graph is a small-world network: highly clustered like a regular graph, yet with small characteristic path length like a random graph.

size, thereby making the time necessary to get equilibrium proportional to log $N$.

The second and third kinds of networks are characterized by the important scale-free condition of links distribution. Formally this means that the distribution density of the number $k$ of links per node, denoted by $P(k)$, has the inverse power law form

$$P(k) \propto \frac{1}{k^{\nu}}. \quad (1)$$

The power index $\nu$ must obey the condition $\nu > 1$ for $P(k)$ to be an acceptable (normalized) distribution density. We note that $\nu = 2$ is the border between the region where the mean number of links per node is finite ($\nu > 2$) and the region where the mean number of links per node is infinite ($\nu < 2$).

The second kind of networks considered in this paper is hierarchical and is illustrated in Figs. 2 and 3. This model shown in Fig. 2, proposed in Ref. [9], generates a scale-free distribution of links with $P(k) \propto 1/k^{\nu}$ with $\nu = \log 3/\log 2 \approx 1.585$.

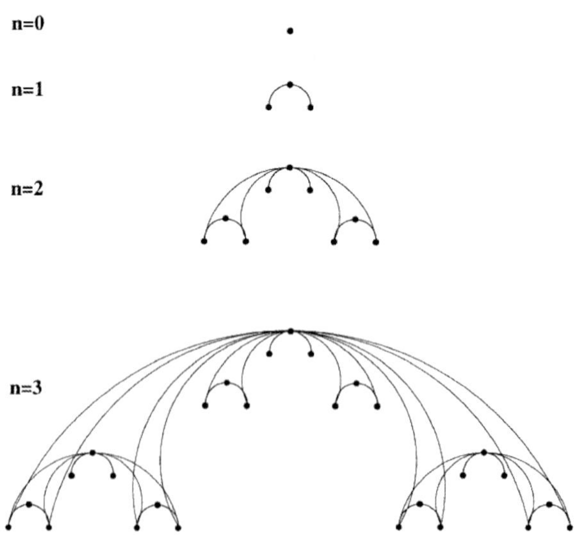


Fig. 2. This figure is adopted from Ref. [9]. Constructing the deterministic scale-free network this figure shows the first four steps of the iterative process.

We shall mainly focus on another network of hierarchical kind, proposed by the authors of Ref. [10], which is closer to the condition of instantaneous equilibrium attainment than the network of Fig. 2. This is therefore a more efficient network and it is illustrated in Fig. 3. By comparing Fig. 3 to Fig. 2 we immediately realize why the latter hierarchical network is more efficient than the former. There exists a hub that is linked to all the nodes of the network, and for this reason a walker moving from even one of poorest nodes can reach any other node in only two steps. The power law index $\nu$ in this case is found numerically to be very close to 1.

Finally, the third class of networks considered in this paper is a scale free network obtained by means of the probabilistic prescription [11]. This method allows us to realize networks with all possible inverse power indexes $\nu > 1$, but the price to pay is that the topological structure of the network remains unknown.

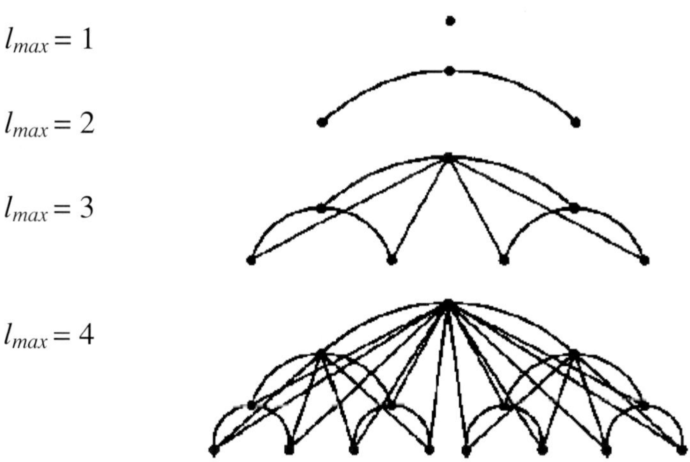


Fig. 3. This figure is adopted from Ref. [10]. Constructing the deterministic scale-free network this figure shows the first four steps of the iterative process.

For the readers to appreciate the main result of this paper, we shortly review the connection between the power index $\mu$ and the network length $L$, established by recent research work. Bollobás and Riordan [12] used rigorous mathematical arguments to derive the results of the model of preferential attachment proposed by Albert and Barabási (AB), [2], and reached the conclusion that the length of the AB network is

$$L = \frac{\log N}{\log(\log N)}. \tag{2}$$

The same value for the length of the preferential attachment network has been more recently derived by Chen *et al.* [13]. In 2003 Cohen and Havlin [14] studied the problem of the relation between $\nu$ and $L$, and found that $\nu = 3$, which is the boundary between the region where $\langle k^2 \rangle < \infty$ ($\nu > 3$) and the region where $\langle k^2 \rangle = \infty$ ($\nu < 3$) has the following length

$$L \approx \frac{\log N}{\log(\log N)}.$$

They also found that for $2 < \nu < 3$ the network length is given by

$$L \approx \log\left(\log N\right). \tag{3}$$

For this reason these authors [14] defined networks as ultra small. In this paper we shall study the condition $\nu < 2$ and for this reason we adopt a different nomenclature, illustrated by Fig. 4. We call the networks with $\nu < 3$ small and ultra small those with $\nu < 2$.

## 3. On the Master Matrix A

The random walk in a network is described by the following master equation

$$\frac{d\rho_i}{dt} = -\rho_i + \sum_{j \neq i} A_{ij} \frac{1}{k_j} \rho_j, \tag{4}$$

for $i$ running from 1 to $N$, where $N$ is the number of network nodes.

Note that $\rho_i(t)$ is the probability for the random walker to be on the site $i$ at time $t$, $k_j$ is the number of links of $j$-th node, and **A** is a matrix whose elements $A_{ij}$ are equal to 1 if the $i$-th node is linked to the $j$-th node, and vanishes otherwise. If we adopt the time integration step $\Delta t = 1$, the master equation of Eq. (4) in a matrix notation becomes

$$\boldsymbol{\rho}(t+1) = \mathbf{B}\boldsymbol{\rho}(t), \tag{5}$$

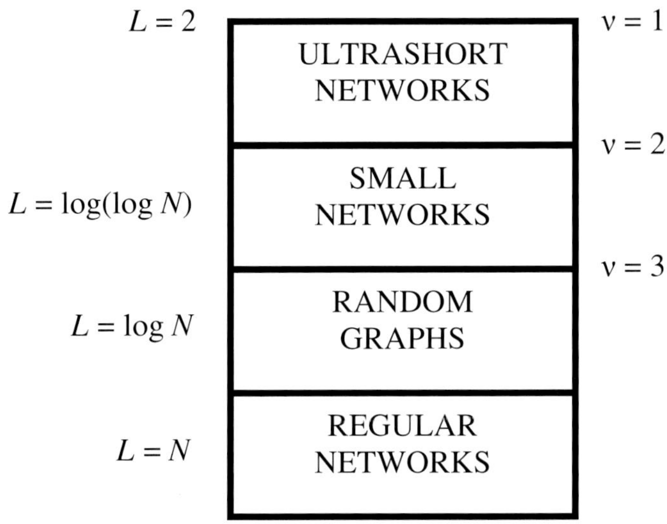


Fig. 4. This figure shows the relation between the free scaling property $\nu$ and the network size $L$.

where the matrix elements $B_{ij}$ are either equal to $1/k_j$ if the $i$-th node is linked to the $j$-th node, or vanish otherwise. It is also convenient to use the following equation

$$\boldsymbol{\rho}(t+1) - \boldsymbol{\rho}(t) = \mathbf{K}\boldsymbol{\rho}(t)\,, \tag{6}$$

where

$$\mathbf{K} \equiv \mathbf{B} - \mathbf{I}\,, \tag{7}$$

and $\mathbf{I}$ denotes the identity matrix. It is straightforward to find the equilibrium distribution predicted by Eq. (6), corresponding to the eigenstate of the matrix $\mathbf{K}$ with vanishing eigenvalue. We have that

$$\rho_i^{(eq)} = \frac{k_i}{\sum_{j=0}^{N} k_j}\,. \tag{8}$$

The eigenvalues of the excited states are negative numbers. Of special interest is the excited eigenvalue with the smallest modulus, insofar as the inverse of this modulus affords an indication of the transition time from an initial out of equilibrium condition to equilibrium. Of course, the actual process of transition to equilibrium depends on the choice of the initial condition.

The condition of almost instantaneous regression to equilibrium makes it easy to evaluate the distribution density of the time distance between two consecutive transits of the random walker through the same node. It is easy to prove that

$$\psi_i(\tau) = \rho_i^{(eq)} \exp\left(-\rho_i^{(eq)}\tau\right). \tag{9}$$

To show this important relation we notice that a random walker that at a give time $t$ is located on a given node $i$, at the time $t+\Delta t$ is found to be on one of the nearest neighbors of the node $i$. Thus the population of node $i$ depends on the trajectory returning on that node. Formally,

$$\rho_i^{(eq)} = \sum_{n=1}^{\infty} \psi_i^{(n)}(t), \tag{10}$$

where $\psi_i^{(n)}(t)$ is the probability that the random walker is on the node $i$ for the $n$th time, exactly at time $t$. The process is renewal, which means that each *step* taken by the random walker is completely independent of any previous steps. As a consequence, each *return* to the node $i$ also has no memory of the earlier returns. We are interested in the long-time limit, and thus the intrinsic renewal nature of the process gives

$$\psi_i^{(n+1)}(t) = \int_0^t dt' \psi_i^{(n)}(t')\psi_i(t-t'). \tag{11}$$

We then set $\psi_i(t) \equiv \psi_i^{(1)}(t)$. By Laplace transforming Eq. (11) and setting $\psi_i^{(0)}(t) = \delta(t)$, we obtain Eq. (9).

## 4. Transition to Equilibrium in Hierarchical Networks

As noted in Section 2 above, the hierarchical network described in Ref. [10] is highly efficient due to its short path length, as shown in Fig. 3. Consequently, equilibrium is reached very quickly in this network. Numerical modeling of this network was performed for a network with 11 levels ($l_{max} = 11$; compare to Fig. 3 above) to determine the time evolution of the population of selected individual nodes.

The results depicted in Figs. 5 and 6 are for the network described in [10] and the case in which many random walkers occupy the top node at time $t = 0$ (i.e., the population of the top node at $t = 0$ is 1). Since this node has the highest degree, equilibrium is reached very quickly. Figure 5 shows the time evolution of this top node; Fig. 6 shows the same for a node in an intermediate level of the network and for a node at the bottom level ($l = 6$ and $l = 11$ respectively). Due to the network topology, the time evolution of the populations is the same for all nodes in the same level when the random walkers all begin at the top node.

Figure 7 depicts results from the other extreme, in which all of the random walkers occupy a single node in the bottom level ($l = 11$) at time $t = 0$. In this case, the time evolution of the populations is not the same for all nodes in the same level because the minimum path lengths from the origin node to nodes in different levels are not the same. As in Figs. 5 and 6, L1 represents the population of the top node, and L11 is the population of the origin node in the bottom level. Data on two intermediate nodes in level 6 ($l = 6$) are provided in Fig. 8. L6min is a node in level 6 that is directly connected to the origin node that the random walker can access in a single step. L6max, however, is on the

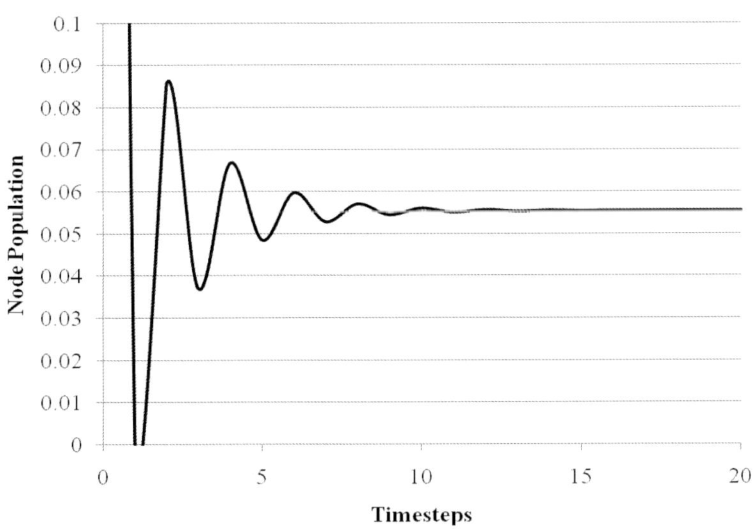


Fig. 5. Population of top node L1.

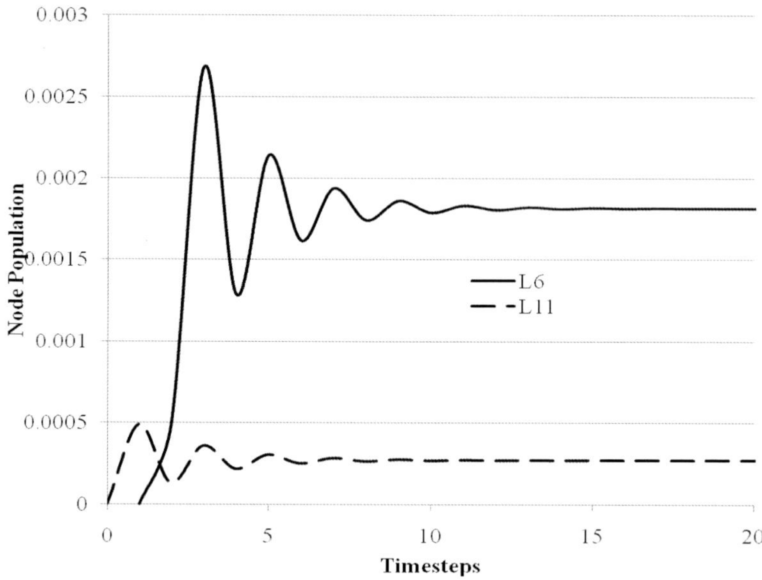


Fig. 6. In this figure, L6 is the population of a node in level 6 of the network ($l = 6$) and L11 is the population of a node in the bottom level ($l = 11$).

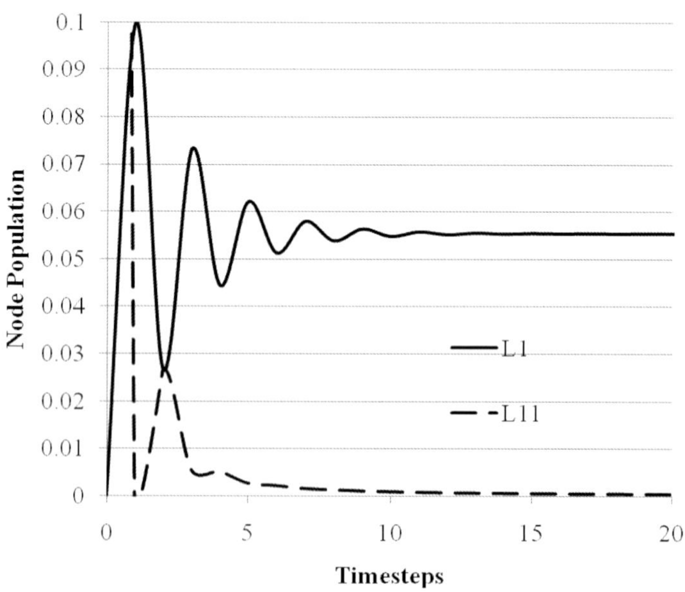


Fig. 7. All random walkers leave from a bottom level node.

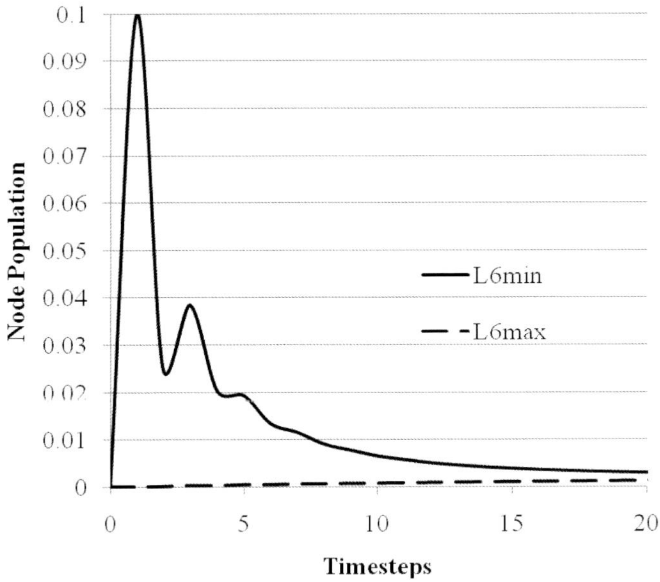


Fig. 8. All random walkers leave from a bottom-level node. These results are for two different intermediate-level nodes.

"far side" of the network: the walker must visit the top node in order to access L6max. (This difference can be more easily visualized by reference to Fig. 3.)

The results show that in both cases, the top node (which is connected to all the other nodes) reaches equilibrium in a number of steps on the order of the number of levels in the network. When all the random walkers start at the top node, the lower level nodes reach equilibrium even faster than the top node. When the walkers start from a lower level in the network, the top node creates a bottleneck, temporarily blocking access to the "other side" of the network and as a result lower level nodes on that side achieve equilibrium more slowly than the lower level nodes on the "origin side" of the network, as evidenced by the L6min and L6max curves in the figures.

Consider now the hierarchical network shown in Fig. 2. When many random walkers are released from a single node in this network, it is observed that the population of each node vanishes for every other time

origin $x = 0$. We now show that the distribution density $\psi(\tau)$ of the times of return to the origin in this case fits the prescription

$$\psi(\tau) \propto \frac{1}{\tau^{\mu}}, \tag{15}$$

with $\mu = 3/2$.

Using the well known property that the solution of Eq. (12) yields $\rho(0,t) \propto 1/t^{0.5}$, the Laplace transform method [i.e., computing the time-dependent Eq. (10) via Eq. (11)] now leads us to Eq. (15) with $\mu = 3/2$. Consequently, we reach the conclusion that the time necessary to get equilibrium, in this case is proportional to $N^2$.

## 5. Return to the Origin in a Scale-free Network

Let us now evaluate the statistical return to the origin $\psi_{stat}(\tau)$ evaluated by running a single trajectory and recording all the time distances between the time of arrival of the random walker to a given node and the immediately earlier time of arrival to the same node. The resulting $\psi_{stat}(\tau)$ is the superposition of the time of return to the same node that is evaluated through the equilibrium superposition of the exponential waiting time distribution densities of Eq. (9).

Following [8] we set, in accordance with Eq. (9),

$$\psi_{stat}(t) = \frac{\sum_i (\rho_i^{(eq)})^2 \exp(-\rho_i^{(eq)} t)}{\sum_i \rho_i^{(eq)}}. \tag{16}$$

### 5.1. *Ad hoc scale-free network*

We define the ad hoc network as a scale-free network that has a merely probabilistic origin [11] and the only available information is that the network is characterized by inverse power law distribution density

$$P(k) \propto \frac{1}{k^{\nu}}, \tag{17}$$

with $\nu > 1$. If we replace the discrete sum of Eq. (16) with a continuous integral and use the assumption of Eq. (17), we get

$$\psi_{stat}(t) \propto \int_{k_{min}}^{\infty} dkP(k)k^2 e^{-kt} \propto \int_{k_{min}}^{\infty} dkk^{2-\nu} e^{-kt} . \tag{18}$$

Substituting the integration variable $y \equiv kt$ yields

$$\psi_{stat}(t) \propto \frac{1}{t^{3-\nu}} F(t) , \tag{19}$$

where

$$F(t) \equiv \int_{k_{min}}^{\infty} dyy^{2-\nu} e^{-y} . \tag{20}$$

We assume that $F(t)$ is a slow function of time, thereby yielding the fundamental result [8]:

$$\psi_{stat}(t) \propto \frac{1}{t^{\mu}} , \tag{21}$$

where

$$\mu = 3 - \nu . \tag{22}$$

The results for a network with $\nu = 1.1$ are illustrated in Fig. 11. The degree distribution for this network is shown in Fig. 12.

### 5.2. *Hierarchical network*

Let us define a hierarchical level $l$, so that $l = 1$ is the president, $l = 2$ are the vice presidents and so on, till to a maximum level $l_{max}$. In this network, higher levels correspond to people with lower importance in the hierarchical scale. The number of nodes at the of level $l$ is

$$n(l) = 2^{l-1} \tag{23}$$

There are thus only one president, two vice-presidents, 4 vice-vice-presidents and so on. The total number of nodes $N$ is

$$N = \sum_{1}^{l_{max}} 2^{l-1} = 2^{l_{max}} - 1 . \tag{24}$$

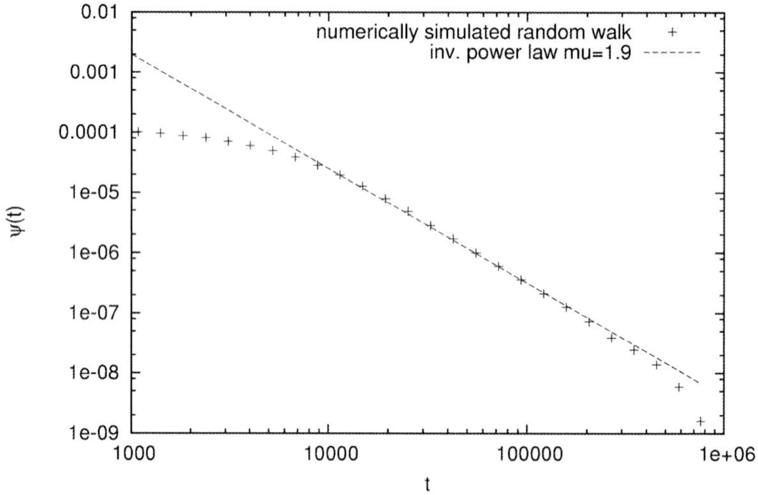


Fig. 11. Waiting time distribution for node returns for the ad hoc scale free network with $\nu = 1.1$. Returns are defined as the number of steps a random walker takes between consecutive visits to a given node.

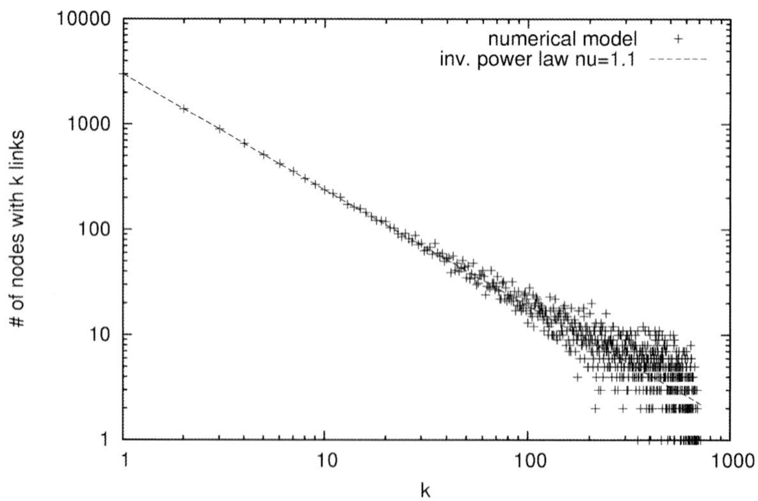


Fig. 12. Degree distribution for the ad hoc network in Fig. 11 along with earlier predictions [11].

Let us now define $k(l)$ as the number of links of a node of level $l$. We evaluate this number by dividing it between ascending nodes $k_a$ and descending nodes $k_d$. It is easy to prove that

$$k_a(l) = l - 1 \tag{25}$$

$$k_d(l) = 2^{l_{max}-l+1} - 2 \tag{26}$$

yielding

$$k(l) = 2^{l_{max}-l+1} + l - 3\,. \tag{27}$$

Equation (25) is trivial, and Eq. (26) is easy to prove. For example the number of links of the president is $N-1$, in agreement with Eq. (27) with $l=1$. In this case all nodes are descending. For a node at a generic level $l$ the descending nodes are just the total number of nodes down his own branch, so the node may act as the president of a network where the maximum level rescales according to the rule $l_{max} \to l_{max} - l$.

To evaluate $n(k)$, the number of nodes having $k$ links, one has to invert Eq. (27) and plug the result into Eq. (23). If one adopts the approximation

$$k(l) \approx 2^{l_{max}-l+1} \tag{28}$$

then it is straightforward to prove that $n(k) \propto k^{-1}$.

Let us now imagine that all the waiting-time distributions for returns have the same form for all the nodes. Therefore the individual waiting times will only depend on $t/\tau_i$, where $\tau_i$ is the mean lifetime of the node $i$. In turn, $\tau_i$ will be inversely proportional to the number of links involving the node $i$, $k_i$. This number, as discussed, only depends on the level $l$ of the node $i$.

It is therefore possible to assume that for each node $i$ at the level $l$

$$\psi_i(t) = ak_i e^{-ak_i t}\,, \tag{29}$$

where $a$ is a constant, defining a time scale. The total waiting time distribution, averaged according to the method illustrated in [7], is

$$\psi(t) = \frac{1}{N}\sum_{l=1}^{l_{max}} 2^{l-1} ak_i e^{-ak_i t}\,, \tag{30}$$

while, averaged according to the method illustrated in [8] is

$$\psi(t) = \frac{\sum_{l=1}^{l_{max}} 2^{l-1} a k_i^2 e^{-a k_i t}}{\sum_{l=1}^{l_{max}} 2^{l-1} k_i} . \tag{31}$$

Figure 13 shows a numerical evaluation of Eq. (31) for $a=1$ and $l_{max}=15$. We notice an exponential cutoff for $t>1/a$, while an inverse-power law is evident for $t<1/a$.

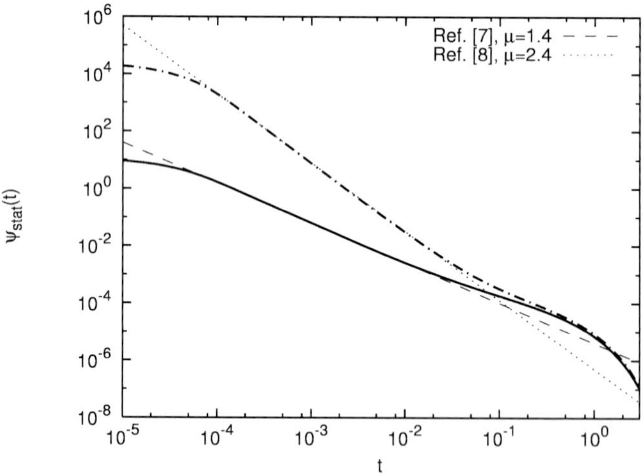


Fig. 13. Plots of Eq. (31) (thick dot-dashed curve) and of Eq. (30) (thick solid curve). The dotted line, fitting the upper curve is an inverse power law with index $\mu=2.4$. The dashed line, fitting the lower curve, is an inverse power law with index $\mu=1.4$. The timescale is in units $1/a$.

## 6. Conclusions

The main results of this paper are:
(i) The regression to equilibrium of a scale-free network with $\nu=1$ is virtually instantaneous.
(ii) The inverse power law of the return times emerges from two opposite conditions, of instantaneous regression to equilibrium and of infinitely slow regression to equilibrium.

(iii) In order to determine which condition applies, it is necessary to start many random walkers from one node, and evaluate the time evolution of the population of the nodes in the network.

**Acknowledgements**

NWH and PG acknowledge financial support from ARO and Welch through grants W911NF-05-1-0205 and B-1577, respectively.

**References**

1. A.L. Barabási, Linked: *How Everything is Connected to Everything Else*, A Plume Book, New York (2003); A. Barrat, M. Barthelemy, A. Vespignani, *Dynamical processes in complex networks*, Cambridge University Press, Cambridge (2008); G. Caldarelli, Scale-Free Networks Oxford University Press, Oxford (2007); S.N. Dorogovtsev and J.F.F. Mendes, *Evolution of Networks: From biological networks to the Internet and WWW*, Oxford University Press, Oxford (2003); M. Newman, A.L. Barabási , D.J. Watts, *The Structure and Dynamics of Networks*, Princeton University Press, Princeton (2006); R. Pastor-Satorras and A. Vespignani, *Evolution and Structure of the Internet: A statistical physics approach*, Cambridge University Press, Cambridge (2004); D.J. Watts, *Six Degrees: The Science of a Connected Age*, W. W. Norton & Company, New York (2003); D.J. Watts, *Small Worlds: The Dynamics of Networks between Order and Randomness*, Princeton University Press, Princeton (1999).
2. R. Albert, A.-L. Barabási, *Statistical mechanics of complex networks*, Rev. Mod. Phys. **74**, 47–97 (2002).
3. V. Latora, M. Marchiori, *Efficient Behavior of Small-World Networks*, Phys. Rev. Lett. **87**, 198701 (1–4) (2001).
4. D.J. Watts, S.H. Strogatz, *Collective dynamics of "small-world" networks*, Nature **393**, 440 (1998).
5. S. Milgram, *The Small World Problem*, Psychology Today, **1**, 60–67 (1967).
6. J. Travers, S, Milgram, *An Experimental Study of the Small World Problem*, Sociometry, **32**, 425–443 (1969).
7. J.D. Noh, H, Rieger, *Random Walks on Complex Networks*, Phys. Rev. Lett. **92**, 118701 (2004).

8. P. Allegrini, L. Palatella, P. Grigolini, *Intermittency and scale-free networks: A dynamical model for human language complexity*, Chaos, Solitons and Fractals **20**, 95–105 (2004).
9. A.-L. Barabási, E. Ravasz, T. Vicsek, *Deterministic scale-free networks*, Physica A **299**, 559 (2001).
10. M. Chen, B. Yu, P. Xu, J. Chen, *A new deterministic complex network model with hierarchical structure*, Physica A **385**, 707 (2007).
11. T. Kalisky, R. Cohen, D. ben-Avraham, S. Havlin, in *Complex Networks*, Lecture Notes in Physics, Springer, Berlin, 650, 3 (2004).
12. B. Bollobas, O. Riordan, *The Diameter of a scale-free random graph*, Combinatorica **24**, 5–34 (2004).
13. F. Chen, Z. Chen, X. Wang, Z. Yuan, *The average path length of scale free networks*, Communication in Nonlinear Science and Numerical Simulation **13**, 1405–1410 (2008).
14. R. Cohen and S. Havlin, *Scale-Free Networks Are Ultrasmall*, Phys. Rev. Lett. **90**, 058701 (2003).

# COHERENCE AND COMPLEXITY

MAURO BOLOGNA
*Instituto de Alta Investigacion, Universidad de Tarapaca,
Casilla 7-D, Arica, Chile*

ELVIS GENESTON
*Department of Physics, La Sierra University,
Riverside, CA 92515, USA*

PAOLO GRIGOLINI
and
MALGORZATA TURALSKA
*Center for Nonlinear Science, University of North Texas,
Denton, Texas 76203-1427, USA*

MIRKO LUKOVIC
*Department of Nonlinear Dynamics & Network Dynamics Group, Max Planck Institute
for Dynamics and Self-Organization, Gottingen 37073, Germany*

We study a model of cooperating units, each of which is a stochastic clock, and we show that when the cooperation strength is critical the model yields coherence and complexity at the same time. We conjecture concerning the relevance of this model to understanding brain dynamics.

## 1. Introduction

The model of cooperating oscillators of Kuramoto [1], which has been vividly illustrated by Strogatz [2], is one of the most popular proposals to describe synchronization. In this Chapter we focus on another model of oscillator synchronization that was originally proposed by Prager *et al.* [3] and has been more recently adopted to discuss the transition to synchronization as an Ising-like phase transition process [4]. A model of the same kind has also been adopted by some authors [5] to discuss the synchronization as a form of athermal phase transition.

$$p_1(\tau) = \frac{1}{3} + \varepsilon_1(\tau)$$
$$p_2(\tau) = \frac{1}{3} + \varepsilon_2(\tau) \tag{9}$$
$$p_3(\tau) = \frac{1}{3} - \varepsilon_1 - \varepsilon_2(\tau),$$

where we have used the condition

$$\varepsilon_3(\tau) = -\varepsilon_1(\tau) - \varepsilon_2(\tau). \tag{10}$$

In fact, if the control parameter $K$ is not large enough to generate synchronization, each of the three states of the stochastic oscillator must host the same number of node. We expect that at the onset of the phase transition to synchronization there will appear a coherent motion of very small amplitude, so that the quantities $\varepsilon_i(t)$ will be much smaller than $1/3$. Thus, by Taylor series expansion of Eq. (7) we obtain up to first order in $\varepsilon_i(t)$,

$$\begin{aligned} &\partial_\tau \varepsilon_1(\tau) + (2-K)\varepsilon_1(\tau) + \varepsilon_2(\tau) = 0 \\ &\partial_\tau \varepsilon_2(\tau) - \varepsilon_1(\tau) + (1-K)\varepsilon_2(\tau) = 0 \end{aligned} \tag{11}$$

The solution of this equation is:

$$\begin{aligned} \varepsilon_1(\tau) &= e^{\left(-\frac{3}{2}+K\right)\tau}\left(c_1 \cos\left[\frac{\sqrt{3}\tau}{2}\right] - \frac{\sqrt{3}}{3}(c_1 + 2c_2)\sin\left[\frac{\sqrt{3}\tau}{2}\right]\right) \\ \varepsilon_2(\tau) &= e^{\left(-\frac{3}{2}+K\right)\tau}\left(c_2 \cos\left[\frac{\sqrt{3}\tau}{2}\right] + \frac{\sqrt{3}}{3}(2c_1 + c_2)\sin\left[\frac{\sqrt{3}\tau}{2}\right]\right). \end{aligned} \tag{12}$$

We see that for $K < 3/2$ the system dynamics decay to the equilibrium condition $p_1 = p_2 = p_3 = 1/3$, and for $K > 3/2$ there is an unphysical exponential increase, due to the fact that the first-order expansion adopted by us to get this result becomes inadequate to study the system dynamics when the control parameter is larger than the critical value

$$K_c = \frac{3}{2}. \tag{13}$$

Another problem generated by the adoption of the first-order perturbation approximation is that the oscillation amplitude $b$ remains unknown.

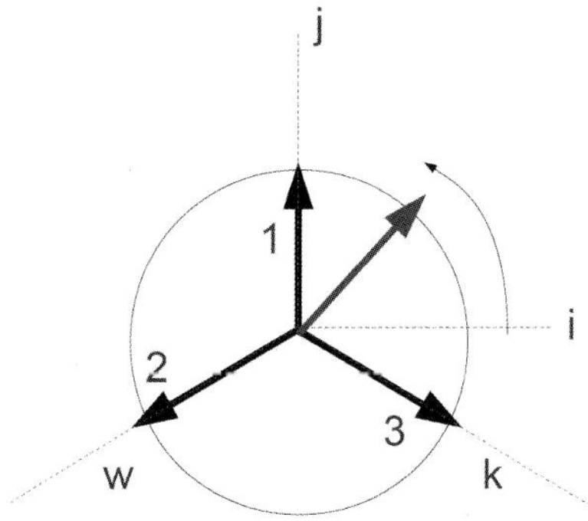


Fig. 1. Illustration of the regular clock generated by the self-organization of infinitely many clocks with the critical value $K_c = 3/2$. It is possible to express this coherent behavior in terms of the probabilities $p_i(t)$ by adopting the non-conventional basis set with the vectors **i**, **k** and **w**, which are not orthogonal. The adoption of the basis set of the orthogonal vectors **i** and **j** is possible provided that the vector **v**(t) is expressed in terms of $\Sigma$ and $\Lambda$ defined in the text.

The theoretical result of Eq. (12) can also be expressed in the equivalent form

$$\varepsilon_1(t) = b \cdot cos(\Omega t), \tag{14}$$

$$\varepsilon_2(t) = b \cdot cos(\Omega t + \frac{2\pi}{3}) \tag{15}$$

and

$$\varepsilon_3(t) = b \cdot cos(\Omega t + \frac{4\pi}{3}), \tag{16}$$

which corresponds to assuming that the hands of the clock at $t = 0$ has the same direction as the versor $j$ of Fig. 1. According to Eq. (12) we have that the frequency $\Omega$ of the global clock is given by

$$\Omega = \frac{\sqrt{3}}{2} g . \tag{17}$$

(see Fig. 2), the amplitude of the harmonic component drops to virtually a vanishing value. Actually, the network's behavior becomes complex [11], the sojourns of the x-component of the velocity vector **v**(t) in the state $v_x = 0$ tends to become infinitely extended in time.

## 3. The Case of a Finite Number of Nodes

Let us analyze now the system dynamics when the number of nodes $N$ is finite. The behavior of the system is illustrated by Fig. 4. We see that after an initial transient from the initial condition where all the oscillators are assumed to be in the same state, the dynamics reach a regime condition that is reminiscent of that of real EEG's. As a consequence of using a finite number of nodes, the system dynamics depart from the regular coherent behavior corresponding to the ideal condition $N = \infty$. Rather than a perennial regular oscillation, we get a sequel of sausages, each of which corresponds to a harmonic oscillation, whose amplitude moves from a minimum value to a maximum value and decreases again to a minimum value. In other words, rather than a regular oscillations of infinite time duration, we have bursts of oscillations, of different size and different length. It is remarkable that the size of these temporary bursts of oscillation can significantly exceeds the amplitude of the regular oscillation generated by the condition $N = \infty$. In Fig. 5 we compare the temporary oscillation bursts to the regular oscillation generated by a

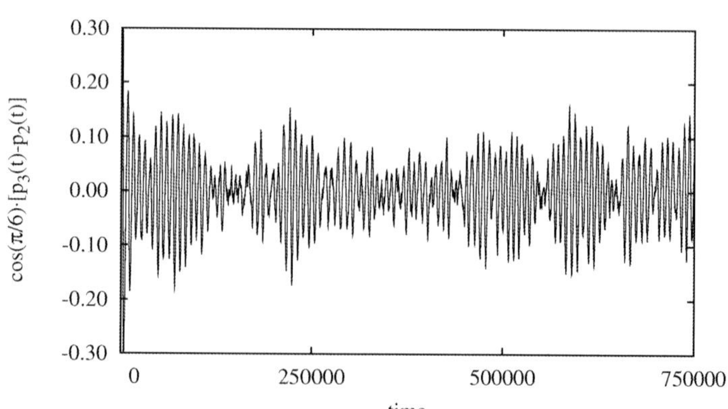


Fig. 4. This is the result of the numerical simulation of the network dynamics with a finite number of nodes at $K = 1.5$.

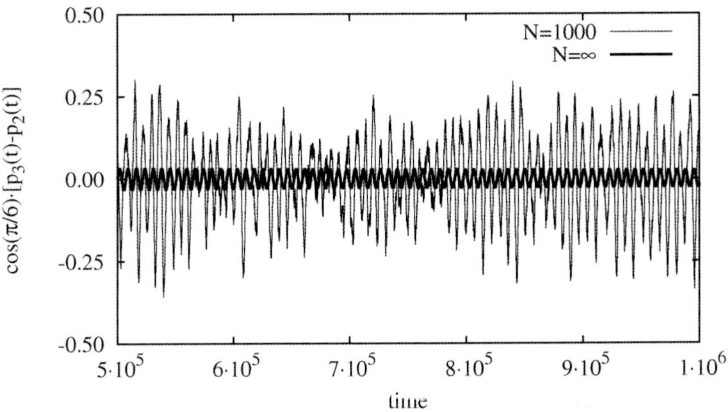


Fig. 5. The fraction of nodes contributing synchronization is much larger than the fraction of nodes contributing synchronization in the ideal condition $N = \infty$. However, this self-organization is temporary, thereby generating a chain of synchronization bursts.

network of infinite size, and we see that it is significantly larger than the amplitude of the regular oscillation. The numerical calculation reveal that the amplitude of the oscillation sausages tends to decrease with increasing $N$, while the length of the sausages becomes larger and larger. We can conclude that the regular oscillation is a sausage whose length became infinitely large.

It is interesting to notice that this behavior can be explained using the theoretical perspective established in earlier work [7]. Turalska *et al.* [7] showed that dynamical complexity requires the action of a finite number of cooperating units. On the one hand the smaller the number of coperating components the larger the amplitude of the mean field created by cooperation before the onset of phase transition, which in the case of infinitely many units should occur at $K = 1.5$. This is equivalent to stating that the smaller the number interacting units the easier to establish the cooperation-induced coherence, which may emerge even at values of $K$ smaller than the critical value. As a consequence, at values of $K$ slighly larger than the critical value the amplitude of the coherent oscillation will turn out to be larger than the theoretical prediction of [11], which is expected to coincide with the result of the numerical solution of Eq. (1) for $N$ very large.

To explain the origin of this interesting phenomenon we have to go back to Eqs. (2), (3) and (4). We see that these transition probabilities depend on the functions $\pi_i(t)$, which fluctuate around the corresponding probabilities $p_i(t)$, due to the fact that $N$ is finite. This is not thermal noise [5], but its effect is similar to that of thermal fluctuations, and it is not only of destructive nature, but it can contribute to increase order, as made evident by the sausages of Fig. 5, whose width is significantly larger than the amplitude of the perennial harmonic signal generated by the condition $N = \infty$.

On the other hand we see that the coherent sausages have a finite time duration, corresponding to the aggregation of the units into clusters of cooperatig units and to their disaggregation. We expect that the length of these coherent sausages will decrease and their time duration will increase with the increase of $N$. It is interesting to notice that the authors of [12] have proposed an interpretation of EEG's signal that seems to be supported by the three-state model of this Chapter. We invite the readers to look at Fig. 8 [12], where the authors [12] suggest that the width of the coherent sausages may be related to the number of interacting neurons, which correspond to the units of the three-state model of this Chapter.

The sequence of oscillation sausages is the manifestation of a process where both complexity and coherence are present. To substantiate this argument, let us focus our attention on the time distance between two consecutive origin re-crossings. The results of this calculation is illustrated by Fig. 6, showing the waiting time distribution density $\psi(\tau)$, where $\tau$ is the time distance between two consecutive re-crossings of the origin. We see that at the time $T = 2\pi / \Omega$ a peak appears, corresponding to the theoretical frequency $\Omega$ of Eq. (17). This peak becomes sharper and sharper with increasing $N$, and it is expected to become infinitely sharp for $N = \infty$. This is the coherence of the title of this paper.

What about network complexity? We would be tempted to claim that the signature of network complexity is evident in the sort-time region of the waiting-time distribution density of Fig. 6, which, by the way, has a clear inverse power-law nature, insofar as it is a straight line in the log-log representation of this figure. However, as we mentioned in the introduction, one of the important requirements to fit the crucial event

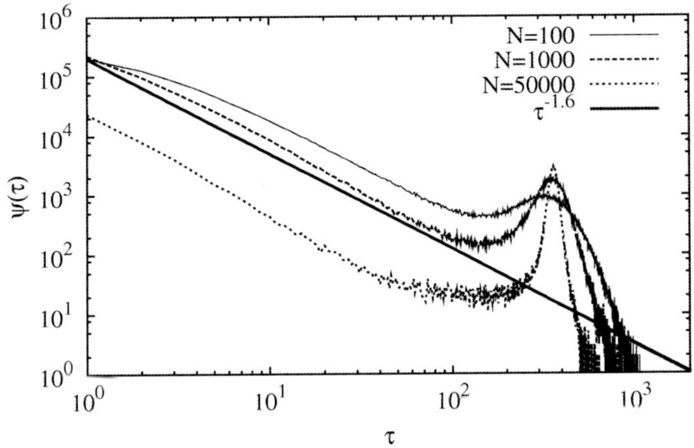


Fig. 6. The waiting-time distribution density for the time intervals between two consecutive recrossing of the origin. The bump at the time $T = 2\pi/\Omega$ becomes sharper and sharper with increasing the number of units N.

condition is the renewal property. To establish if the dynamical process under study hosts renewal events or not we make the renewal aging experiment of Ref. [13].

The results are illustrated in Figs. 7 and 8. In Fig. 7 we see that the brand new survival probability undergoes an abrupt drop to zero at a time scale of the order of $T = 2\pi / \Omega$, thereby signaling the coherent nature of this process. As explained [13,6] the renewal nature of the process is established by evaluating the aged waiting time distribution. The aged waiting-time distribution density is determined by starting the waiting time process at a distance $t_a$ from an earlier recrossing. Usually, the corresponding survival probability yields a slower decay, as shown by both dashed and thin solid curve in the short-time region. However, in the case of the model of this Chapter this traditional property of crucial events is lost at a time distance of the order of 100, where the aged survival probabilities show a decay faster than the decay of the brand new survival probability. Note that the dashed curve indicates how the survival probability of the system ages, and the thin solid curve shows how it should age in the renewal case. We see that the dashed curve departs from the thin solid line, thereby signaling that the process is not renewal.

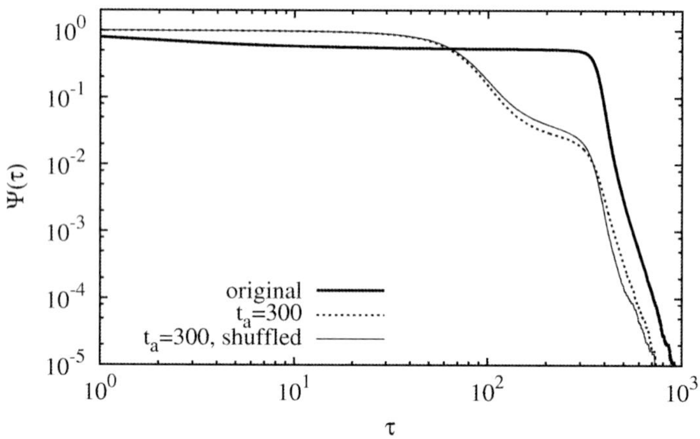


Fig. 7. Origin re-crossings. The curves of this figure illustrate the probability that up to a given time distance $t$ from a given origin recrossing there is no further origin recrossing. The red curve is the ordinary survival probability. We see that it drops abruptly to zero at the time $T = 2\pi / \Omega$, corresponding to the system's coherence. The dashed and the thin solid curve denotes the aged survival probabilities, corresponding to the probability that no recrossing event occurs at a given distance $t$ from the beginning of the waiting process that is done at a time distance $t_a$ from an earlier re-crossing.

The breakdown of the renewal condition is made even more evident by noticing that the age $t_a$ of the aged survival probabilities of Fig. 7 is smaller than the period $T$ of the coherent process. In Fig. 8 we set $t_a = T$, and we find that as a consequence of this resonance condition, the deviation from the renewal state extends from the long to the short times. In conclusion, the origin re-crossing is not a good indicator of the real complexity of the system.

We now show that a reliable signature of the system complexity is given by the distribution of the time length of the coherence sausages. Fig. 9 shows that the coherence bump disappears and that the waiting time distribution density is characterized by an inverse power law index $\mu$ of the order of 1.55. The condition

$$\mu < 2 \tag{20}$$

corresponds to a complexity condition that is expected to be a signature of biological processes, a clear sign of their out of equilibrium nature [14].

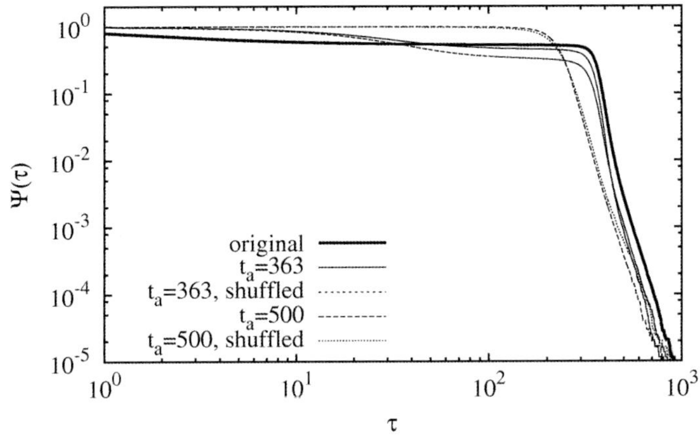


Fig. 8. The same as in Fig. 7, with $t_a = 2\pi / \Omega$ . This a resonant condition where the age $t_a$ is tuned to the system's coherence.

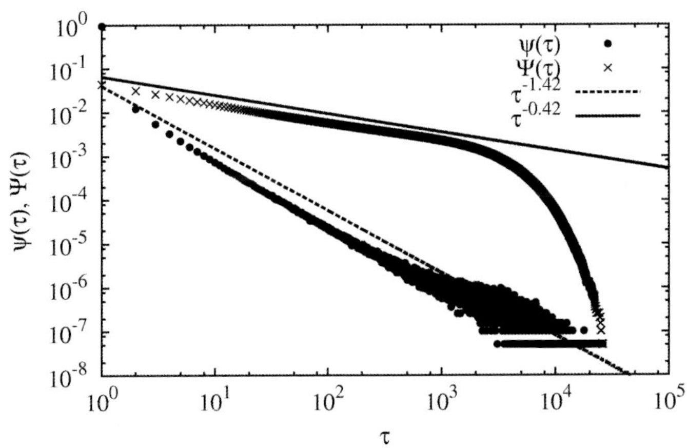


Fig. 9. Sausage length. In this figure we plot the probability that a sausage has a time length $t$ , and the corresponding survival probability, namely the probability that an intermittent coherent process is still on up a time $t$ .

## 4. Concluding Remarks

In his very interesting report about the Kuramoto model [2] Strogatz pointed out that one of issues still to be settled with the Kuramoto model is its behavior for a finite number of nodes $N$ . One of the attractive properties that Strogatz and his co-workers found in later work is the existence of chimera states [15]. In short, the model of Kuramoto makes the synchronized oscillators emerge from a sea of unsynchronized units that are called *chimera states*. Is there a connection between the chimera states and the complexity of the model of the present article?

We think that a connection certainly exists with the unsettled issues of the Kuramoto model and a finite number of nodes $N$ [2]. In the present Chapter we have shown that the condition of a finite number of interacting units is a source of time complexity similar to that discovered in earlier work [7]. The ideal condition $N = \infty$ generates a perennial coherence. The authors of Ref. [4] proved that this condition of perennial coherence can also be realized without adopting the all-to-all coupling condition of this Chapter, if the units are the nodes of a regular Euclidean lattice, interacting only with the nearest neighbors, and the Euclidean dimension $d$ is equal to 2, or larger. Herein we established that the not yet fully explored condition of a finite number of nodes may host the surprising condition that the time complexity [7] coexists with coherence. We emphasize that our chain of coherent sausages is generated by a process of re-organization where clusters of synchronized oscillators are destroyed and replaced by other units emerging from the sea of non-synchronized states. The remarkable similarity of our chain of coherent sausages with the real EEG [12] suggests the ambitious purpose of adopting this mixture of coherence and complexity to shed lights into the brain function.

**Acknowledgements**

MB, EG, PG and MT acknowledge financial support from ARO and Welch through grants W911NF-05-1-0205 and B-1577, respectively.

## References

1. Y. Kuramoto, in: H. Arakai (Ed.), International Symposium on Mathematical Problems in Theoretical Physics, Lecture Notes in Physics, Vol. 39, Springer, New York, 1975, p. 420; *Chemical Oscillations, Waves, and Turbulence*, Springer, Berlin, 1984; Progr. Theoret. Phys. Suppl. **79**, 223 (1984).

2. S. Strogatz, *From Kuramoto to Crawford: exploring the onset of synchronization in populations of coupled oscillators*, Physica D **143**, 1 (2000).
3. T. Prager, B. Naundorf, L. Schimansky-Geier, Physica A **325**, 176 (2003).
4. K. Wood, C. Van den Broeck, R. Kawai, K. Lindenberg, *Universality of Synchrony: Critical Behavior in a Discrete Model of Stochastic Phase-Coupled Oscillators*, Phys. Rev. Lett. **96**, 145701 (2006); *Critical Behavior and synchronization of discrete stochastic phase-coupled oscillators*, Phys. Rev. E **74**, 03113 (2006); *Effects of disorder on synchronization of discrete phase-coupled oscillators*, Phys. Rev. E **75**, 041107 (2007); *Continuous and discontinuous phase transitions and partial synchronization in stochastic three-state oscillators*, Phys. Rev. E **76**, 041132 (2007).
5. B. Fernandez, L.S. Tsimring, *Athermal Dynamics of Strongly Coupled Stochastic Three-State Oscillators* Phys. Rev. Lett. **100**, 165705 (2008).
6. S. Bianco, E. Geneston, P. Grigolini, M. Ignaccolo, "Renewal aging as emerging property of phase synchronization", Physica A **387**, 1387 (2008).
7. M. Turalska, M. Lukovic, B.J. West, P. Grigolini, "Complexity and synchronization", Phys. Rev. E **80**, 021110 (2009).
8. M. Turalska, E. Geneston, P. Grigolini, B.J. West, "Dynamical origin of topological complexity", Phys. Rev. Lett. under review (2010).
9. F. Barbi, M. Bologna, P. Grigolini, *Linear response to perturbation of nonexponential renewal processes*, Phys. Rev. Lett. **95**, 220601(1-4) (2005); P. Allegrini, M. Bologna, P. Grigolini, B.J. West, *Fluctuation-dissipation theorem for event-dominated processes*, Phys. Rev. Lett. **99**, 010603 (2007); G. Aquino, P. Grigolini, B.J. West, *Linear response and Fluctuation-Dissipation Theorem for non-Poissonian renewal processes* Europhys. Lett. **80**, 10002 (2007); L. Silvestri, L. Fronzoni, P. Grigolini, P. Allegrini, *Event-driven power-law relaxation in weak turbulence*, Phys. Rev. Lett. **102**, 014502 (2009); P. Allegrini, M. Bologna, L. Fronzoni, P. Grigolini, L. Silvestri, *Experimental quenching of harmonic stimuli: Universality of linear response theory*, Phys. Rev. Lett. **103**, 030602 (2009).
10. G. Aquino, M. Bologna, P. Grigolini, B.J. West, *Beyond the Death of Linear Response: 1/f optimal information transport*, Phys. Rev. Lett., under review.
11. M. Bologna *et al.*, work in progress.
12. A.A. Fingelkurt, A.A. Fingelkurts, C.F.M. Neves, *Natural world physical, brain operational, and mind phenomenal space-time*, Physics of Life Reviews **7**, 195–249 (2010).

13. S. Bianco, P. Grigolini, P. Paradisi, *Fluorescence intermittency in blinking quantum dots: Renewal or slow modulation?* J. Chem. Phys. **123**, 174704 (2005).
14. B.J. West, E. Geneston, P. Grigolini, *Maximizing information exchange between complex networks*, Phys. Rep. **468**, 1 (2008).
15. D.M. Abrams, S.H. Strogatz, *Chimera States for Coupled Oscillators*, Phys. Rev. Lett. **93**, 174102 (2004); D.M. Abrams, R. Mirollo, S.H. Strogatz, D.A. Wiley, *Solvable Model for Chimera States of Coupled Oscillators*, Phys. Rev. Lett. **101**, 084103 (2008); E.A. Martens, C.R. Laing, S.H. Strogatz, *Solvable Model of Spiral Wave Chimeras*, Phys. Rev. Lett. **104**, 044101 (2010).

# QUAKES IN COMPLEX SYSTEMS AS A SIGNATURE OF COOPERATION

ELVIS GENESTON

*Department of Physics, La Sierra University,*
*4500 Riverwalk Parkway, Riverside, CA 92515*

PAOLO GRIGOLINI

*Center for Nonlinear Science, University of North Texas,*
*P.O. Box 311427, Denton, Texas 76203-1427, USA*

In this Chapter we study a set of cooperating neurons generating quakes with the same size distribution as the neuronal avalanches in mature cultured networks, recently revealed by the experimental observation. We prove that in addition to this form of complexity this model yields a form of phase transition generating also temporal complexity. This means that the distance from two consecutive quakes fits the prescription of a Mittag-Leffler (ML) function renewal theory. There exists a critical value of the cooperation parameter at which this description applies to the whole time regime. By increasing the cooperation parameter the ML theory breaks down and the sequence of quakes tend to become periodic and the quakes have the same intensity. We make the conjecture that the recently discovered principle of complexity matching may be conveniently applied to the neuro-physiological processes that are properly described by this model.

## 1. Introduction

Many complex processes are characterized by abrupt configuration changes, or quakes, whose signature has been established in cell assemblies,[1] brain activity,[2] spin glasses,[3] fluorescence bursts induced by enzyme fluctuations[4] and Blinking Quantum Dots (BQD),[5] to name only a few. The discovery that the Deborah number of the BQD diverges[5] is forcing some researchers[6] to go beyond one of the pillars of statistical physics: the ergodic hypothesis. The discovery that there are quakes in the music composition and that these quakes match the brain quakes[7] suggests that the transport of information between two complex networks may rest on the dialogue of the respective quakes, and that this is made possible by new forms of fluctuation-dissipation processes.[8]

Although the occurrence of avalanches in complex systems is a widely recognized property, there is still some confusion on the concept of criticality. As far as neural dynamics are concerned, the concept of criticality has been closely connected to the theory of self-organized criticality.[9] This has led to the very exciting discovery of avalanches in neural dynamics[10,11] with the experimental detection of the crucial scaling coefficient $q = 1.5$, whose theoretical derivation was found in a later work by Zapperi *et al.*[12] On the other hand, more recently, the authors of Ref. 13 found that the Ising model at phase transition reproduces very well the correlation structure revealed by the FMRI analysis of the brain resting data. In a recent review paper[14] Chialvo illustrates these very stimulating results as a manifestation of the brain criticality without discussing how one form of criticality is connected to the other. The main aim of the Chapter is to illustrate a neural model generating avalanches with $q = 1.5$ in correspondence of a form of phase transition at criticality.

## 2. Mittag-Leffler Function: Detection of the Phase-transition Occurrence

The Mittag-Leffler (ML) function[15] is the mathematical tool that we use in this Chapter to prove the emergence of cooperation from a model of interacting neurons. To facilitate the efforts that the readers should do to understand our demonstration, it is convenient to review some important aspects of the ML function. In the literature of complex systems there is a general agreement that relaxation departs from the conventional exponential decay, but no essential agreement on the specific form of relaxation to use to properly fit the experimental results.

The researchers working in this field are divided into two main groups. The first group assigns to the survival probability $\Psi(t)$ the following analytical form

$$\Psi(t) = exp\left(-(\lambda t)^{\alpha}\right), \qquad (1)$$

with $\alpha < 1$ and $\lambda > 0$ being the parameter defining the time scale of the process. Of course, setting $\alpha = 1$ would make it possible for us to recover the traditional form of exponential relaxation. The function of Eq. (1) is usually termed *stretched exponential function*. Many authors refer to this as the Kohlrausch–Williams–Watts (KWW) function, from the names of the authors who originally proposed this fitting form.[16]

The authors of the second group adopt for their fitting purposes the inverse power law form:

$$\Psi(t) = \frac{1}{(1+\lambda t)^{\alpha}}. \tag{2}$$

Millonas and Ray[17] proposed a model where the system of interest is coupled to an out of equilibrium thermal bath, which is responsible for making the experimentally observed relaxation depart from the ordinary form of exponential decay. These authors find a form of non-exponential relaxation that differ "from the power law and stretched exponential as little as the two differ from each other".[17]

More recently, Metzler and Klafter[18] presented a generalization of the classical exponential relaxation based on the fractional Fokker-Planck equation framework that naturally led them to assign to $\Psi(t)$ the form of the ML function. The ML function has a simple analytical form in the Laplace representation. In this Chapter we adopt throughout for the Laplace transform of the function $f(t)$ the notation

$$\hat{f}(u) \equiv \int_0^{\infty} d\tau f(\tau) exp(-u\tau). \tag{3}$$

Using this notation, if we assign to $\Psi(t)$ the ML form, we have

$$\hat{\Psi}(u) = \frac{1}{u + \lambda^{\alpha} u^{1-\alpha}}. \tag{4}$$

Via inverse Laplace transformation we get

$$\Psi(t) = E_{\alpha}\left(-(\lambda t)^{\alpha}\right), \tag{5}$$

where $E_{\alpha}(z)$ is the ML function expressed with respect to time. The series expansion of $E_{\alpha}(z)$ is

$$E_{\alpha}(z) = \sum_{n=0}^{\infty} \frac{(-z)^n}{\Gamma(1+\alpha n)}. \tag{6}$$

For $0 < \alpha \leq 1$, the ML function is positive and strictly monotonically decreasing.[15]

It is important to stress that in the short-time regime $t < 1/\lambda$ the ML function has the form of the stretched exponential function of Eq. (1). Although there is a general agreement that complex systems manifest their deviation from the predictions of ordinary statistical physics in the long-time region with fat tails, in practice the quantitative evaluation of complexity in the long-time regime can turn out to either difficult or impossible.

The authors of Refs. 7 and 19 adopted the form

$$\hat{\Psi}(u) = \frac{1}{u + \lambda^{\alpha}(u + \Gamma_t)^{1-\alpha}}, \tag{7}$$

with $\alpha < 1$. In the case $\Gamma_t = 0$ this is the Laplace transform of the ML function of Eq. (4). However, the experimental observation has forced the authors of Refs. 7 and 19 to introduce the parameter $\Gamma_t > 0$. The experimental results are explained by setting $1/\Gamma_t$ of the order of $1/\lambda$ and both much larger than the unit time step: in this case Eq. (4) generates a function $\Psi(t)$ indistinguishable from a stretched exponential function. In the case of the earlier work of Refs. 7 and 19 the ML complexity is violated in the large time limit by environmental fluctuations that make the system fall in the exponential regime. In the case of this Chapter, on the contrary, the breakdown of the ML condition is due to the same cooperative effect as that making the ML function emerge in the first place. We shall see, in fact, that in the model studied in this Chapter the single units are neurons. Each neuron moves towards a threshold where it fires. When one neuron fires, all the other neurons make a step ahead (towards threshold) by the quantity $k$, which is the cooperation parameter. When the parameter is very large, all the neurons fire at the same time, and the process become exactly periodic. The ML picture is invalidated when this periodic condition becomes significant.

The critical condition emerging from the numerical calculations illustrated in this Chapter is a physical property of the ML regime and corresponds to the transition from $\alpha = 1$ to $\alpha = 0.77$ representing the cooperation-induced complexity. In the Ising model[20] the control parameter is given by temperature, and the effect of the phase transition is the global magnetization that at criticality begins departing from the vanishing value that is realized in the absence of cooperation where the single spins get randomly and with the same probability either the "up" or the "down" condition.

In the case of this Chapter, we observe quakes, namely events corresponding to one or more neurons firing. The physical meaning of the survival probability $\Psi(t)$ is defined by referring ourselves to the occurrence of a quake as time origin. Given that a quake occurs at the time origin, what is the probability that no new quake occurs up to time t? The probability of no quake occurrence is the survival probability $\Psi(t)$. In the absence of cooperation

$$\Psi(t) = exp(-Gt). \tag{8}$$

In this case $\alpha = 1$ and $\lambda = G$. The quantity corresponding to the magnetization of the Ising model is

$$m = \lambda^{\alpha} - G. \tag{9}$$

At criticality $\lambda^{\alpha} - G$ begins departing from the vanishing value. In practice, we observe numerically the quantity

$$g(k) = \lambda(k)^{\alpha(k)}, \tag{10}$$

and we use it to establish the critical value of $k$ at which phase transition occurs.

## 3. Description of the Model

Cooperation among many different units is considered to be the crucial ingredient for the emergence of intelligence perceived as a form of large distance correlation. The birds of a swarm are located at different positions of an Euclidean subset and interact via local couplings. This local interaction is reminiscent of the condition realized by the well known two-dimensional Ising model,[20] where each unit interacts with only four nearest neighbors. However, when the control parameter is sufficiently intense a phase transition occurs[21] which establishes the coordinated behavior of the whole swarm. This form of phase transition has been more recently observed in real flocks of birds[22] and the authors of this very interesting paper have emphasized the similarity between the long-range correlation emerging from the local interaction of birds and the ferromagnetic phase transition that at criticality generated the maximal correlation among units. This is in fact closely connected to the phase transition criticality mentioned by Chialvo[14] and observed in the brain by the authors of Ref. 13. However, it is not clear how this form of criticality is related to the avalanches of Refs. 10 and 11. To establish this connection we have to realize cooperation in a different way that seems to be more appropriate to describe the interaction among firing neurons. For this reason the model that we plan to discuss in this Chapter is almost identical to that recently used by Levina *et al.*,[23] the main difference being that we use only excitatory interaction and do not allow the inhibitory synapses of Ref. 23 to act. The work of Ref. 23 is an interesting example of self-organized criticality applied to neuronal dynamics. We are convinced that this model generates also a form of phase transition similar to the ferromagnetic and this is the interesting aspect that this Chapter aims to point out.

The cooperative agents of this Chapter are $N$ neurons and a complex quake is the simultaneous firing of two or more neurons. With no cooperation, the neuron potential $x(t)$ is described by the following equation[49]

$$x(t+1) = (1-\gamma)x(t) + S + \sigma\xi(t), \tag{11}$$

where $t$ is a natural number, $\xi(t)$ is a variable getting either the value of 1 or $-1$, with equal probability, with no memory of the earlier values, and $\gamma \ll 1$ so as to make the integer time virtually continuous when $\gamma t \approx 1$. The quantity $\sigma$ is the noise intensity. The variable $x$ moves from the initial condition $x = 0$ and $S > 0$ favors its motion towards the threshold $x = 1$. At the threshold, the neuron fires and resets back to the initial value $x = 0$. When $\sigma = 0$, it is straightforward to prove that the time necessary for the neuron to reach the threshold is given by

$$T_{MS} = \frac{1}{\gamma} ln\left(\frac{1}{1-\frac{\gamma}{S}}\right). \tag{12}$$

We assume, with Mirollo and Strogatz,[24] that the neurons cooperate, namely, that when one neuron fires all the other neurons make a step ahead by the quantity $k < 1$ which is the cooperation strength.

This condition is equivalent to assume that each neuron interacts with all the other neurons, and it will be referred to as All-To-All (ATA) coupling condition.

Mirollo and Strogatz (MS) found the exact solution of this problem when $\sigma = 0$[24] and proved that, after a transient, all the neurons fire at the same time, with time period $T_{MS}$ of Eq. (12), the MS time.

As an effect of noise it may take a time either shorter or longer than $T_{MS}$ for a neuron to fire. Furthermore, the time distance between two consecutive firings may be much shorter than $T_{MS}$. When $k = 0$, even if we assign to all the neurons the same initial condition, $x = 0$, due to stochastic fluctuations the neurons fire at different times so that the system eventually reaches a steady state with a constant firing rate $G$ given by

$$G = \frac{N}{<\tau>}, \tag{13}$$

where $<\tau>$ denotes the mean time between two consecutive firings of the single neuron. For $\sigma \ll 1$, $<\tau> \approx T_{MS}$. From the condition of constant rate $G$ we immediately derive the Poisson survival probability

$$\Psi(\tau) = exp(-G\tau). \tag{14}$$

We shall see hereby that this heuristic argument agrees very well with numerical results. For the numerical simulation we select the condition

$$G \ll 1 \ll N \ll T_{MS}, \tag{15}$$

yielding

$$\frac{1}{G} \approx \frac{T_{MS}}{N} \ll T_{MS}, \tag{16}$$

thereby realizing the time scale of interest for this Chapter, which is much shorter than the MS time. It is evident that the condition of non-interacting neurons fits Eq. (4) with $\alpha = 1$ and

$$\lambda(k = 0) = G. \tag{17}$$

For our numerical calculations we adopt the values $\sigma = 0.001$, $N = 100$, $\gamma = 0.0001$ and $S = 0.00019$. We have assessed numerically that $\langle \tau \rangle = 7431$ thereby yielding, thanks to Eq. (13), $G = 0.0135$, and $T_{MS} = 7472.14$. It is important to point out that for larger values of $N$ the same properties are observed as long as the condition (15) is applied.

In the case of no interaction the probability that two neurons fire at the same time is $G^2$, thereby making the simultaneous firing of two neurons virtually impossible, due to $G \ll 1$ (see Eq. (15)). The cooperation among neurons, on the contrary, creates events corresponding to the simultaneous firing of two or more neurons. As a consequence, the decay of the survival probability becomes slower. In fact, when cooperation is so strong as to subside the stochastic fluctuation, and all the neurons fire at the same time, $\Psi(t)$ remains constant up to $t = T_{MS}$.

The numerical results of Fig. 1 and Fig. 2 afford a compelling evidence of these properties. Curve 1 of Fig. 1, corresponding to the case of no interaction, coincides, as emphasized by the insert of this figure, with the exponential function of Eq. (8), thereby confirming that in the case of this model Poisson statistics correspond to the condition of no cooperation. Curve 5 of Fig. 1, corresponding to $k = 0.1$, is very close to the ideal condition where all the neurons fire at the same time, as predicted by Mirollo and Strogatz[24] in the case $\sigma = 0$.

The readers can convince themselves that this is indeed the case by observing Fig. 2. We see that although the neurons do not fire exactly at the same time the firing process occurs in burst and the width of these bursts for $k \to \infty$ is expected to become infinitely small so as to produce sequences identical to the periodic delta-like functions of Mirollo and Strogatz.[24]

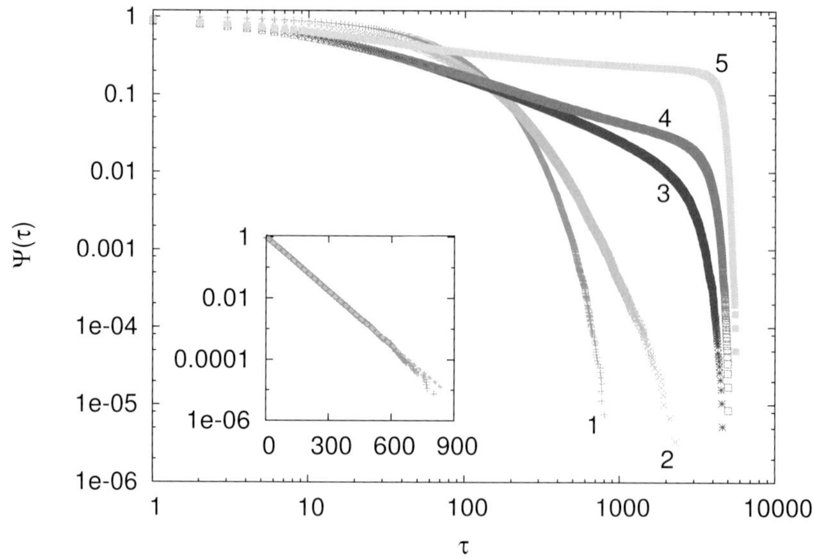


Fig. 1. The survival probability $\Psi(\tau)$ as a function of $\tau$. The vertical arrows denote $1/G = 74.31$ (left) and $T_{MS} = 7472.14$ (right). Curves 1, 2, 3, 4 and 5 refer to $k = 0, 0.002, 0.005, 0.01, 0.1$, respectively. The insert shows curve 1 in a log-linear representation; the dashed straight line is Eq. (14) with G = 0.0135.

## 4. Subordination Theory

The theoretical perspective to interpret the numerical results of this Chapter is afforded by the *subordination theory*. We do not afford an exhaustive review of this theory rooting on the Continuous Time Random Walk (CTRW) of Montroll and Weiss.[25] The subordination theory adopted in this Chapter is the same as that used in the earlier work of Ref. 19 and illustrated in detail in the review paper of Ref. 26. For earlier work on subordination we refer the reader to Refs. 27 and 28.

For the sake of reader's convenience we give here a concise description of the subordination theory adapted to the specific problem under study in this Chapter.

The parameter $g < 1$ is the probability that at a given time, referred to as natural time,[26] one or more neurons fire. The probability that no firing takes place is $1 - g$. The probability that no firing takes place up to the natural time $n$ is

$$\Psi^{(NF)}(n) = (1-g)^n = exp\left(ln(1-g)^n\right) = exp\left(-nln\left(\frac{1}{1-g}\right)\right). \quad (18)$$

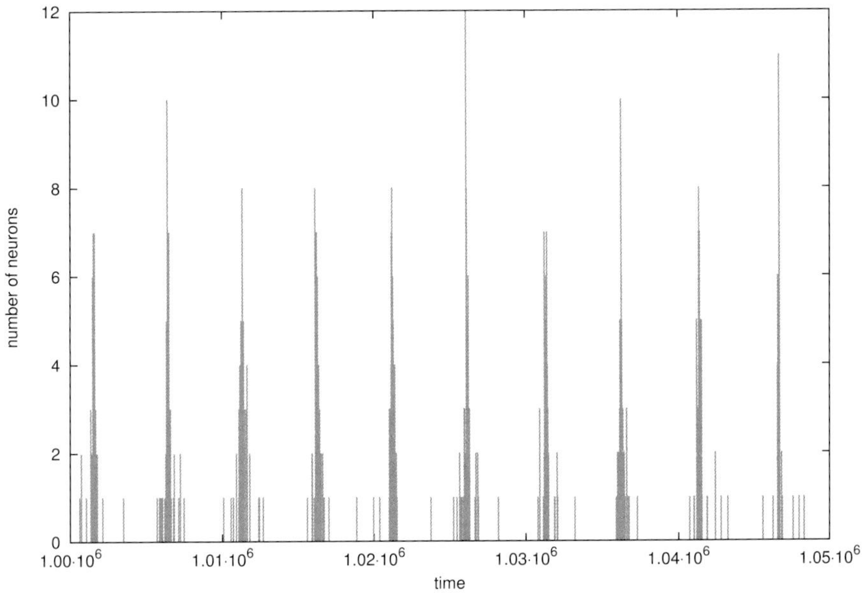


Fig. 2. Number of neuron firings as a function of time for $k = 0.01$.

We see that in the natural time dominion the survival probability is an ordinary exponential function. Note that we can interpret $g$ as

$$g = \frac{n_f}{N}, \tag{19}$$

where $n_f$ is the number of firing neurons and $N$, as earlier stated, is the total number of neurons.

It is evident that as a result of cooperation the probability $g$ must increase. At the same time there should be time regions where no firing occurs. In the case of perfect cooperation $g = 1$, but there are time intervals of length $T_{MS}$ where no firing occurs.

To establish a more realistic representation of the effect of cooperation we must define the function $\psi^{(S)}(t)$. This is the waiting time distribution density defining the probability of finding a time interval of length $t$, where not only no firing occurs, but not even the condition happens where the firing is possible.

As an effect of cooperation at criticality the function $\psi(t)$ makes a transition from the exponential to the inverse power law form.

The numerical calculations of this Chapter allow us to evaluate the survival probability $\Psi(t)$ in the real continuous time. This is given by the formula

$$\Psi(t) = \sum_{n=0}^{\infty} \int_0^t \psi_n^{(S)}(t')(1-g)^n \Psi^{(S)}(t-t')dt', \tag{20}$$

where

$$\Psi^{(S)}(t) = \int_t^{\infty} \psi^{(S)}(t')dt'. \tag{21}$$

The meaning of the very important formula of Eq. (20) is the following. The function $\Psi(t)$ establishes that no firing occurs up to the time distance $t$ from the last firing event. We have take into account how many times a firing event became possible. This is given by a generic number $n$ ranging from 0 to $\infty$. In practice, it is impossible that infinitely many favorable conditions occur in a finite time. However, we have to take into account that the continuous time $t$ must be interpreted as a sort of macroscopic time, which is in fact so large as to ideally accommodate infinitely many potential events.

To make it possible for the reader to fully understand the meaning of Eq. (20) we have to introduce the concept of renewal. In fact, a very important property of the subordination theory is the renewal nature of the potential events. The renewal condition allows us to define properly the function $\psi_n^{(S)}(t)$ and to establish an attractive way of evaluating it. The function $\psi_n^{(S)}(t)$ is the probability density for the occurrence of $n$ potential firings, the last of which occurs at time $t$, whereas the others occur at earlier times. The renewal property allows us to express $\psi_n^{(S)}(t)$ as a function of $\psi_{n-1}^{(S)}(t)$ as follows

$$\psi_n^{(S)}(t) = \int_0^t \psi_{n-1}^{(S)}(t')\psi_1^{(S)}(t-t')dt', \tag{22}$$

where $\psi_1^{(S)}(t) \equiv \psi^{(S)}(t-t')$. The important fact that we are considering the last event as the origin of time is taken into account by setting

$$\psi_0^{(S)} \equiv \delta(t). \tag{23}$$

Using both Eq. (22) and Eq. (23) we find that the Laplace transform of $\psi_n^{(S)}(t)$ is

$$\hat{\psi}_n(u) = \hat{\psi}(u)^n. \tag{24}$$

At this stage we can get a full understanding of Eq. (20). The quantity $\psi_n^{(S)}(t')$ is the probability that $n$ potential events occurred, the last of which took place at time $t' < t$. This makes it possible to weight the probability than no firing took place, $(1-g)^n$. We have to fill the gap between the time $t'$ and the time $t$ at which the survival probability $\Psi(t)$ is evaluated. Therefore we must complete the weight by the factor $\Psi^{(S)}(t-t')$, which is the probability that no potential firing occurs between the last potential firing at time $t'$ and the observation time $t$. The Laplace transform of the function $\Psi^{(S)}(t)$ of Eq. (21) is defined by

$$\hat{\Psi}^{(S)}(u) = \frac{1}{u}\left(1 - \hat{\psi}^{(S)}(u)\right). \tag{25}$$

The time convolution nature of Eq. (20) makes it easy to evaluate the Laplace transform of $\Psi(t)$. Using Eq. (24) and Eq. (25) we obtain

$$\hat{\Psi}(u) = \left[\sum_{n=0}^{\infty}\left(\hat{\psi}^{(S)}(u)(1-g)\right)^n\right]\frac{1}{u}\left(1 - \hat{\psi}^{(S)}(u)\right). \tag{26}$$

Summing the geometric series after some easy algebra we get the fundamental result

$$\hat{\Psi}(u) = \frac{1}{u + g\hat{\Phi}(u)}, \tag{27}$$

where

$$\hat{\Phi}(u) \equiv \frac{u\hat{\psi}^{(S)}(u)}{1 - \hat{\psi}^{(S)}(u)}. \tag{28}$$

It is important to stress that the fundamental result of Eq. (27) not only allows to find the critical phase-transition condition that it is the main result of this Chapter. Eq. (27) allows us to explore also the condition of perfect synchronization that is reached at $k = \infty$. In fact, in this case $g = 1$ and

$$\psi^{(S)}(t) = \delta(t - T_{MS}). \tag{29}$$

Thus,

$$\hat{\psi}^{(S)}(u) = e^{-uT_{MS}}, \tag{30}$$

$$\hat{\Psi}(u) = \frac{ue^{-uT_{MS}}}{1 - e^{-uT_{MS}}} \tag{31}$$

and

$$\hat{\Psi}(u) = \frac{1 - e^{-uT_{MS}}}{u}, \tag{32}$$

which is the Laplace transform of a survival probability holding the value of 1 up to the Mirollo and Strogatz time, at which it instantaneously drops to 0.

evaluate the crucial order parameter of Eq. (10). This figure has a remarkable similarity with the typical phase transition of the Ising model when the number of units is not infinite. In the ideal case of a phase transition with infinite units, the order parameter of Eq. (9) should vanish up to the critical value $k_c$ and for $k > k_c$ should have the behavior

$$m(k) = constant(k - k_c)^{\beta}, \tag{33}$$

where $\beta$ is an unknown power index, thereby making $m(k)$ a function with discontinuous derivative. In the case of a finite number of units the function $m(k)$ becomes continuous thereby making uncertain the value of $k$ at which the phase transition occurs. In the case of the cooperative model of this Chapter no theory is currently available to predict the critical value $k_c$. Therefore we plot $g(k)$ rather than $m(k)$, but we estimate that the critical value of the control parameter should be $k_c = 0.002$.

At criticality a significant change from Poisson statistics to ML statistics occur. At $k < k_c$, the parameter $g << 1$, and the function $\psi^{(S)}(t)$ has the exponential form

$$\psi^{(S)}(t) = re^{-rt}. \tag{34}$$

Using the fundamental result of Eq. (27) we get the exponential form of Eq. (14) with

$$G = rg. \tag{35}$$

It is reasonable to set $r = 1$ in accordance with the fact that in the absence of cooperation the potential firing events occur at each time step. It is convenient to let the reader notice that in the numerical treatment of this Chapter the elementary time step is $\Delta t = 1$. Thus we find, for $k < k_c$,

$$g(k) = G, \tag{36}$$

in accordance with curve 1 of Fig. 1.

At $k = k_c$, the function $\psi^{(S)}(t)$ makes an abrupt change from the exponential form of Eq. (34) with $r = 1$ to the inverse power law form

$$\psi^{(S)}(t) = (\mu_S - 1)\frac{T_S^{\mu_S - 1}}{(t + T_S)^{\mu_S}}, \tag{37}$$

which defines $T_S$ as the microscopic time scale, at which the inverse power law does not operate yet. In the non-ergodic case $\mu_S < 2$, using the Laplace transform method of Ref. 29, we obtain that the limiting condition $u \to 0$ yields (4) with

$$\lambda = \left[ \frac{g}{T_S^\alpha \Gamma(1-\alpha)} \right]^{1/\alpha}, \tag{38}$$

where $\Gamma(1-\alpha)$ is the Gamma function, with

$$\alpha = \mu_S - 1. \tag{39}$$

When $g = 1$, Eq. (27) yields

$$\Psi(t) = \Psi^{(S)}(t). \tag{40}$$

Note that the theoretical interpretation that we are adopting to explain the results of the numerical simulation of this Chapter is based on the distinction between *action* and *event*. When $g = 1$ each action yields an event and the experimental $\Psi(t)$ directly inherits the power law index $\mu_S$. When $g < 1$ some actions do not produce events. When $g \ll 1$ the majority of actions do not produce events and the emergence of the power law index $\mu_S$ is delayed. However, a careful analysis of the short-time stretched exponential allows us to reveal the complexity index $\mu_S$, even when the inverse power law portion of the ML function is not visible due to some truncation process either due to the environment randomness or to the cooperation-induced synchronization. We use the symbol $\mu$ rather than $\mu_S$ when the complexity of $\psi^{(S)}(t)$ become experimentally visible.

There are infinitely many $\psi^{(S)}(\tau)'s$ realizing this condition. Let us restrict our analysis to the $\psi^{(S)}(\tau)'s$ with $T_S$ defined by $T_S^\alpha \Gamma(1-\alpha) = C$, with $C$ being a constant that for simplicity's sake we set equal to 1 for the same reasons as those adopted to set $r = 1$ in Eq. (35). In conclusion, we obtain Eq. (10) which is the order parameter that we use in this Chapter to establish the phase transition emerging from the neuron cooperation.

Assuming that $k_c = 0.002$, we conclude that at criticality the transition from the function $\psi^{(S)}(t)$ of Eq. (34) to the function $\psi^{(S)}(t)$ of Eq. (37) occurs. The ideal order parameter $m(k)$ is a continuous function, even though its derivative is discontinuous. Thus we can safely set

$$G = g(k_c) = \lambda(k_c)^{\alpha_c}. \tag{41}$$

To shed light into this interesting form of phase transition let us observe Fig. 3. This important figure corresponds to the log-log representation of curve 2 of Fig. 1, which, in turn, is the survival probability $\Psi(t)$ emerging at criticality, according to our approximated estimation of $k_c$. This representation leads to an accurate determination of $\alpha$ that turns out to be

$\alpha = 0.77$. The stretched exponential of the ML function contains already information on the form of temporal complexity that is expected to emerge at larger time under the form of the inverse power law (see Eq. (2))

$$\Psi(t) \propto \frac{1}{t^{\mu-1}}. \tag{42}$$

We conclude that at criticality the time distance between two consecutive quakes is given by the waiting time distribution

$$\psi(t) \propto \frac{1}{t^{\mu}}, \tag{43}$$

with

$$\mu = 1.77 < 2. \tag{44}$$

We notice that although difficult to realize, due to the smooth nature of $g(k)$ at criticality, an abrupt change takes place in this region. Using Eq. (41) we find that the stretched exponential emerging at criticality has the value

$$\lambda \approx 0.004, \tag{45}$$

which is one order of magnitude smaller than the value of $\lambda$ appearing at $k = 0.01$, as shown by Fig. 4.

In conclusion, at criticality the ML picture seems to be a very accurate description of the system's complexity. In Fig. 3 we note that at about $t = 1000 << T_{MS} = 7472.14$ a deviation from the straight line appears. This does not have anything to do with the breakdown of the ML picture. It rather corresponds to the transition from the stretched exponential regime to the inverse power law regime. Due to the condition of Eq. (15) at $t = T_{MS}$ the survival probability is so small as to make it impossible for us to perceive the limitation that the Mirollo-Strogatz periodicity sets to the renewal theory that we adopt to interpret the numerical results of this Chapter. With increasing values of $k$ the decay of $\Psi(t)$ becomes slower thereby making the sign of synchronization perceptible. This is especially evident in Fig. 4, corresponding to $k = 0.01$. On the basis of the numerical results of Fig. 4 we can state that the ML picture is a very good picture for the complexity of the model studied in this Chapter for $k$ ranging from $k_c \approx 0.002$ to $k_{bd} \approx 0.01$.

For $k > k_{bd}$ the ML picture is broken and with it the renewal picture that we have adopted to interpret the numerical results of this Chapter. From $k_{bd}$ on a new condition appears where the synchronization properties of the model of Mirollo and Strogatz become evident.

With respect to the earlier work of Refs. 7,19,30 the numerical results of this Chapter show that the inverse power law regime is limited by the effects of cooperation itself rather than from the influence of a random environment. In other words, cooperation generates temporal complexity but an excess of cooperation yields the breakdown of the regime of temporal complexity.

As a final remark in this Section, we would like to notice that the abrupt transition from $\alpha = 1$ to $\alpha = 0.77$ allows us to bypass a problem that would affect Eq. (38) with a treatment with no discontinuity. In fact, the Gamma function diverges for $\alpha \to 1$. Setting the condition that $T_S^{\alpha}\Gamma(1-\alpha)$ is constant would force the microscopic time $T_S$ vanish.

## 6. Temporal Complexity at Criticality

To appreciate the importance of temporal complexity, we have to make a short excursion on the subject of phase transition, whose importance for complex processes is acknowledged by an increasing number of scientists, with little attention, however, to the condition of temporal complexity emerging at criticality.

Phase transitions and critical phenomena occur frequently in nature and have been widely studied by physicists, see for instance.[31] The Ising model[32] originally introduced to explain ferromagnetic phase transition is well known, and the exact solution found by Onsager[33] for the occurrence of phase transition in the two-dimensional case is widely recognized as an example of outstanding theoretical achievement. In the last few years some scientists used the Ising model to shed light into biological and neurophysiological processes.[13,14,34–36] More precisely, the authors of Ref. 34 used the Ising model to explain the collective behavior of biological networks and the authors of[13,14,35] adopted the Ising model for the main purpose of supporting their conviction that the brain works at criticality without establishing a clear distinction between phase transition and self-organized criticality.[9] Finally, the authors of Ref. 36 used the Ising model to explore the cognition properties of the brain.

Couzin[37] made the attractive conjecture that the cooperative interaction between the birds of a flock may mimic the brain cognition properties, and the experimental observation of Cavagna *et al.*[22] confirmed the criticality condition of a coordinated flock of birds. It is important to stress that the connection between phase transition processes and cooperative behavior of biological systems was already stressed in the pioneer paper by

Viczek *et al.*[21] It is thought[34] that biological systems in general operate at criticality.

The phase transition condition is not confined to biological systems. The recent work on econo-physics[38,39] suggests that the same arguments can be applied to sociological systems. As a matter of fact, Sornette[39] proposes a model of cooperative economical interaction, of the same type as the Ising model, which generates at criticality a transition from subcritical to supercritical patterns very similar to those produced by the Ising model. As stressed by the authors of Ref. 13, the Ising patterns at criticality correspond to the emergence of a correlation links yielding a scale-free network statistically indistinguishable from that experimentally observed on the brain, thanks to functional magnetic resonance imaging.

These interesting studies emphasized the spatial and network complexity emerging from the cooperative interaction of the system's units, but overlooked the temporal complexity of these systems. In this paper we are making an attempt to fill this gap and we plan to prove the temporal complexity emerging at criticality. The temporal complexity is defined as follows. The time dynamics of the complex systems is characterized by the occurrence of significant events, which may be the financial crashes of Ref. 39, the brain quakes of Refs. 13,14 or the changes of direction of a flock of birds.[21,22] The time distance between two consecutive events is given by a distribution density $\psi(t)$, which, drastically departing from the conventional Poisson statistics, get the inverse power law form:

$$\psi(\tau) \propto \frac{1}{\tau^{\mu}}, \tag{46}$$

with $\mu < 2$. This is nothing but Eq. (43) that we rewrite here to attract readers's attention. The occurrence of an event does not have any memory of the occurrence of the earlier events. This property is usually denoted as renewal, but it must not be confused with the ordinary Poisson and Markov condition: the signal generated by these events is characterized by long-range correlation in time, and, most importantly, it is essentially non stationary, thereby breaking the ergodicity that is a fundamental property of statistical physics.[40] We refer to it as non-Poisson renewal condition. The significant event is interpreted as a failure whose occurrence brings back the system to the brand new condition. We refer to these events as *crucial events*.

Quite surprisingly, in the literature there is only a limited number of authors[41–43] who realized the connection between phase transition and temporal complexity. An exhaustive and accurate discussion of this property can be found in the work of Ref. 44. Turalska *et al.*[44] prove that at criticality the inverse power law regime becomes infinitely extended with increasing the number of cooperating units.

It has to be pointed out that the work of Refs. 41–44 established a connection between temporal complexity and phase transition only in the specific case of Ising-like phase transition processes, thereby leaving unexplored the case of self-organized criticality. It is beyond the purpose of this Chapter to discuss to what extent the model here under study can be interpreted as an example of self-organized criticality. We notice however that Chialvo in his recent discussion on the criticality of neuron dynamics[14] stresses the importance of criticality with no concern about the lack of a clear connection between the criticality of Ising-like phase transition and the criticality of self-organized criticality. The model studied in this paper yields a form of criticality similar to that of the Ising model and at same time generates a cascade of quakes with the crucial exponent close to 3/2 that is frequently interpreted as a manifestation of self-organized criticality. This aspect will be discussed in the next Section.

## 7. Avalanches

In this Section we focus our attention on Fig. 6. This figure depicts the size distribution $D(s)$ of neuronal avalanches emerging from the numerical calculations of this Chapter. The similarity with the experimental results of Ref.[10] is impressive. Both the numerical results of Fig. 6 and the experimental results of Ref. 10 obey the prescription

$$D(s) \propto \frac{1}{s^q}, \tag{47}$$

with

$$q = \frac{3}{2}, \tag{48}$$

followed by a fast exponential drop. The crucial exponent of Eq. (48) coincides with the theoretical predictions of.[12] The fast exponential drop seems to be an obvious consequence of the finite number of neurons used in this Chapter and of the finite number of electrodes of the experimental work of Ref. 10. In, fact, we see that the exponential drop occurs at higher and higher values of $s$ with increasing the number of electrodes.[10] Fig. 6 suggests that although the crucial power index $q = 1.5$ emerges as an effect

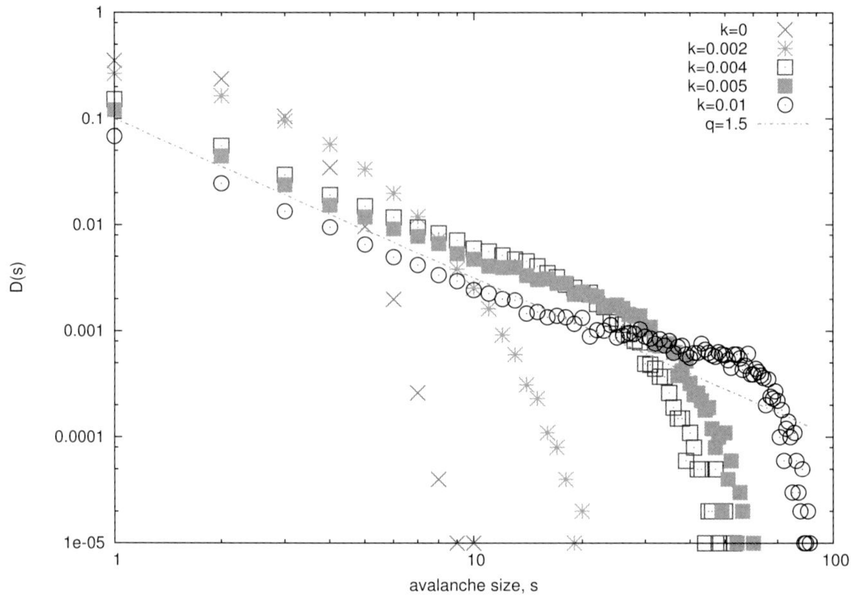


Fig. 6. Avalanche size distribution of selected values of $k$ from Fig. 5 using time window $\Delta t = 100$. The points for $k = 0$ show an exponential distribution, a stretched exponential for $k = 0.002$, and a power law distribution for $k = 0.01$.

of neuron cooperation, it does not seem to be a special property of criticality. In fact, although the condition $k = 0.005$, corresponding to the blue curve of Fig. 6 is very far from criticality, its slope is not very different from $q = 1.5$. However, it is remarkable that at $k = 0.002$ a transition occurs from the exponential to the stretched exponential regime.

## 8. Complex Network

We have seen that the complex quakes are a sign of cooperation. In this Chapter we focus on the neuron cooperation. However, it is plausible that also human activity with social cooperation may produce the same deviation from the exponential behavior as recently stressed by Barabasi.[30] In light of the remarks done in the earlier sections, we conclude that the most important issue to address is whether the model adopted generates an Ising-like phase transition or whether a cascade of complex quakes emerges from the model.

In this Section we plan to discuss shortly a different issue. The calculations discussed so far refer to the ATA condition. What about the case where the ATA condition does not apply and a firing neuron influences only the neurons linked to it?

We consider four different networks. The first one is a regular (R) network, with the sites distributed on a circle with each site coupled to its $2m_{nn}$ nearest neighbors. The other three networks are complex: the Watts and Strogatz (WS) network characterized by high clustering;[45] the Barabási and Albert (BA) network,[46] which is known to be scale free and finally the Holme and Kim (HK) network,[47] which shares with WS the high clustering and with BA the scale free condition. We assign to all three complex networks the same mean number of links, $m_l$, and we set $m_l = 2m_{nn}$.

The motivation for the numerical calculations of this Section are given by recent studies[45,46] going beyond the traditional concept of random networks and establishing two new classes of networks, with high clustering[45] and power law distribution of edges.[46] More recently a new type of network has been proposed[47] with both high clustering and power law distribution of edges. It has been noticed that the scale free property of the complex networks is shared by the human brain.[48]

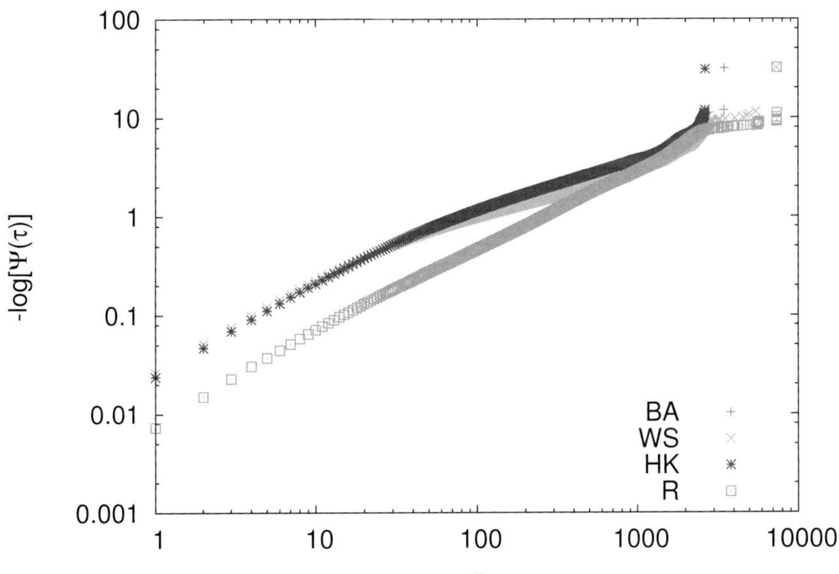


Fig. 7. The function $\Psi(\tau)$ as a function of $\tau$. We adopt the condition $m_l = 10$. The bottom line corresponds to the condition of a regular network. The other three networks, WS, AB and HK, merge into to top curve. Parameters are $\gamma = 0.0001$, $S = 0.00019$, $\sigma = 5 \times 10^{-5}$ and k = 0.3.

The results are illustrated in Fig. 7. The log-log representation of $-log[\Psi(\tau)]$ as a function of $\tau$ emphasizes the emergence of the stretched exponential condition for the R network. We see that for the same value of $k$ the complex networks yield a pronounced transition to the inverse power law behavior, thereby proving that they favor the emergence of the regime of high synchronization,[50] which may become excessive[51] and unhealthy.[52] Note that releasing the ATA condition requires larger values of $k$ for the actions to become visible, $k = 0.3$ of Fig. 5 rather than $k = 0.01$ of Fig. 1.

## 9. Concluding Remarks

We note that Barabasi[30] stressed the emergence of the inverse power law behavior, properly truncated, as a consequence of the cooperative nature of human actions. Apparently, the numerical results of this Chapter set a limit to the generality of this view, insofar as we see that the numerically monitored events suggest the ML behavior rather than the inverse power law prescription to be the general consequence of cooperation. Wihin the ML perspective the stretched exponential seems to play a role more important than the inverse power law. Actually, this Chapter shows that the emergence of the stretched exponential is a consequence of the fact that not all the actions generate events. If we focus our attention on actions more than on events the main effect of cooperation is again the birth of a waiting time distribution density with the predominant character of an inverse power law.

The cooperative actions have a limited success, and only from time to time, an action corresponds to a quake. The rare quakes are described by the ML function of Eq. (4) and the actions are virtually invisible. The complex networks make actions more successful and consequently visible. However, a further increase of cooperation may break Eq. (4) and generate a time periodic process.

The authors of Ref. 11 stressed the remarkable stability of the temporal patterns revealed by their experimental work and made the conjecture that this may shed light into the origin of epilepsy. We therefore make the conjecture that the healthy condition of the brain may correspond to the onset of phase transition, where, as we have seen, the periodic nature of the process is negligible.

The results of this Chapter are essentially based on computer simulation, but they suggest that a general phase transition theory may be found in a model that generates at the same time quakes with the distribution size predicted by SOC. This model is a natural extension of that originally proposed by Mirollo and Strogatz[24] and it is remarkable that this model generates a sequence of quakes that obey the theoretical predictions based on SOC.[12]

It is also remarkable that this model generates a kind of phase transition that is similar to the well known phase transition generated by the Ising model. The temporal complexity emerging from the criticality of Ising-like model has been recently proved with compelling numerical evidence in the work of Ref. 44. To the best of our knowledge, this Chapter affords the first numerical demonstration that temporal complexity emerges also from a model producing SOC avalanches.

We have to remark that an important new principle has been recently proved. This is the principle of Complexity Matching.[53] According to this principle a complex system is sensitive only to the influence exerted on it by a stimulus with the same complexity. Thus, regardless of whether the brain is conveniently described by either the Ising-like models of Ref. 13 or by the cooperative model of this Chapter, it is expected to respond to stimuli with the same complexity. Of course, the most attractive form of stimulus is given by music.[7] This brings us back to the concept of music thought of as the mirror of mind. Should the model discussed in this Chapter be relevant as a model of the brain dynamics, we should focus our attention on music compositions generating quakes with the same complexity as the brain.

## Acknowledgments

E.G. and P.G. acknowledge financial support from ARO and Welch through grants W911NF-05-1-0205 and B-1577, respectively.

## References

1. D. Plenz and T.C. Thiagarajan, Trends in Neurosciences **30**, 101 (2007).
2. A. Ya. Kaplan *et al.*, Signal Processing **85**, 2190 (2005).
3. P. Sibani, Eur. Phys. J B **58**, 483 (2007).
4. X. Michalet *et al.*, Chem. Rev. **106**, 1785 (2006).
5. X. Brokmann *et al.*, Phys. Rev. Lett. **90**, 120601 (2003); G. Margolin and E. Barkai, *ibid.* **94**, 080601 (2005), G. Margolin *et al.*, Adv. Chem. Phys. **133**, 327 (2006).
6. A. Rebenshtock and E. Barkai, Phys. Rev. Lett. **99**, 210601 (2007).
7. S. Bianco, M. Ignaccolo, M.S. Rider, M.J. Ross, P. Winsor, P. Grigolini, *Brain, music, and non-Poisson renewal processes* **75**, 061911(2007).

8. P. Allegrini *et al.*, Phys. Rev. Lett. **99**, 010603 (2007); G. Aquino *et al.*, Europhys. Lett. **80**, 10002 (2007).
9. P. Bak, *How Nature Works* (Oxford Univ. Press, 1997).
10. J.M. Beggs, D. Plenz, *Neuronal avalanches in neocortical circuits*, J. Neurosci. **23**, 11167-11177 (2003).
11. J.M. Beggs, D. Plenz, *Neuronal avalanches are diverse and precise activity patterns that are stable for many hours in cortical slice cultures*, J. Neurosci. **24**, 5216–5229 (2004).
12. S. Zapperi, L.K. Baekgaard, H.E. Stanley, *Self-organized branching processes: meanfield theory for avalanches*, Phys. Rev. Lett. **75**, 4071–4074 (1995).
13. D. Fraiman, P. Balenzuela, J. Foss, D.R. Chialvo, *Ising-like dynamics in large-scale functional brain networks*, Phys. Rev. E **79**, 061922 (2009).
14. D.R. Chialvo, *Emergent complex neural dynamics*, Nature Physics **6**, 744 (2010).
15. G.M. Mittag-Leffler, C. R. Acad. Sci. Paris **137**, 554 (1903); R. Acad. dei Lincei, Rendiconti **13**, 3 (1904); Acta Math. **29**, 101 (1905).
16. R. Kohlrausch, Ann. Phys. (Leipzig) **12**, 393 (1847); G. Williams, D.C. Watts, Trans. Faraday Soc. **66**, 80 (1970).
17. M.M. Millonas, C. Ray, *Relaxation Kineics of Nonlinear Systems Coupled to a Nonequilibrium Bath*, Phys. Rev. Lett. **75**, 1110 (1995).
18. R. Metzler, J. Klafter, *From stretched exponential to inverse power-law: fractional dynamics, Cole-Cole relaxation processes, and beyond*, Journal of Non-Crystalline Solids **305**, 81(2002).
19. R. Failla, P. Grigolini, M. Ignaccolo, A. Schwettmann, *Random growth of interfaces as a subordinated process*, Phys. Rev E **70**, 010101 (2004).
20. J.P. Sethna, *Statistical Mechanics: Entropy, Order Parameters, and Complexity* Clarendon Press-Oxford (2010).
21. T. Vicsek, A. Czirk, E. Ben-Jacob, I., Cohen, O. Shochet, *Novel type of phase transition in a system of self-driven particles* Phys. Rev. Lett. **75**, 1226–1229 (1995).
22. A. Cavagna, A. Cimarelli, I. Giardina, G. Parisi, R. Santagati, F. Stefanini, M. Viale, *Scale-free correlations in starling flocks* Proceedings of the National Academy of Sciences of the United States of America 107 (26), pp. 11865–11870 (2010).
23. A. Levina, J.M. Herrmann, T. Geisel, *Dynamical synapses causing self-organized criticality in neural networks* Nature Physics **3**, 857–860 (2007).
24. R.E. Mirollo and S.H. Strogatz, *Synchronization of pulse-coupled biological oscillators*, SIAM J. Appl. Math. **50**, 1645–1662 (1990).
25. E.W. Montroll and G.H. Weiss, *Random Walks on Lattices. II*, J. Math. Phys. **6**, 167–181 (1965).
26. B.J. West, E.L. Geneston, P. Grigolini, *Maximizing information exchange between complex networks*, Phys. Rep. **468**, 1–99 (2008).
27. I.M. Sokolov, Phys. Rev. A **63**, 011104 (2000); E. Barkai and R.J. Silbey, J. Phys. Chem. B (**104**), 3866 (2000); R. Metzler and J. Klafter, J. Phys. Chem. B **104**, 3851 (2000).

28. I.M. Sokolov, J. Klafter, Chaos **15**, 026103 (2005); R. Gorenflo, F. Mainardi and A. Vivoli, Chaos, Solitons and Fractals **34**, 87 (2007).
29. B.J. West, M. Bologna, P. Grigolini, *Physics of Fractal Operators*, Springer-Verlag, New York (2003).
30. A.-L. Barabási, Nature **207**, 435 (2005).
31. H.E. Stanley, *Introduction to Phase Transition and Critical Phenomena*, Oxford University Press, New York (1971).
32. E. Ising, (1925), *Beitrag zur Theorie des Ferromagnetismus*, Z. Phys. **31**, 253–258 (1925).
33. L. Onsager, *Crystal statistics. I. A two-dimensional model with an order-disorder transition*, Phys. Rev. 65, 117–149 (1944).
34. T. Mora, W. Bialek, *Are biological systems poised at criticality?* arXiv: 1012.2242v1[q-bio.QM] 10 December 2010.
35. M.G. Kitzbichler, M.L. Smith, S.R. Christensen, E. Bullmore, *Broadband Criticality of Human Brain Network Synchronization* PLoS Computational Biology **5**, 1–13 (2009).
36. W. Truccolo, L.R. Hochberg, J.P. Donoghue, Nature Neuroscience **13**, 105–113 (2010).
37. I. Couzin, *Collective minds*, Nature **445**, 715 (2007).
38. H.E. Stanley, V. Plerou, Quantitative Finance, *Scaling and universality in economics: empirical results and theoretical interpretation* **1**, 563–567 (2001).
39. D. Sornette,*Critical market crashes*, Phys. Rep. **378**, 1–99 (2003).
40. A. Rebenshtok, E. Barkai, *Weakly Non-Ergodic Statistical Physics* J. Stat. Phys. **133**, 565–586 (2008).
41. Y.F. Contoyiannis, F.K. Diakonos, *Criticality and intermittency in the order parameter space* Phys. Lett. A **268**, 286–292 (2000).
42. S. Bianco, E. Geneston, P. Grigolini, M. Ignaccolo, *Renewal aging as emerging property of phase synchronization* Physica A: Statistical Mechanics and its Applications **387**, 1387–1392 (2008).
43. M. Turalska, M. Lukovic, B.J. West, P. Grigolini, *Complexity and Synchronization*, Phys. Rev. E **80**, 021110 (2009).
44. M. Turalska, B.J. West, P. Grigolini, *Temporal complexity at the phase transition*, submitted to Phys. Rev. E.
45. D.J. Watts and S.H. Strogatz, *Collective dynamics of small-world networks*, Nature (London) **393**, 440–442 (1998).
46. A.-L. Barabási and R. Albert, *Emergence of scaling in random networks*, Science **286**, 509–512 (1999).
47. Petter Holme* and Beom Jun Kim, *Growing scale-free networks with tunable clustering*, Phys. Rev. E **65**, 026107, 1–4 (2002).
48. B.J. Kim, *Geographical Coarse Graining of Complex Networks* Phys. Rev. Lett. **93**, 168701, 1–4 (2004).
49. H.C. Tuckwell, *Introduction to Theoretical Neurobiology* Volume 2, Cambridge University Press, Cambridge (1988).
50. Curve 4 of Fig. 1 suggests that the inverse power law regime is the close precursor of the regime of excessive synchronization, thereby implying that complex networks may favor the emergence of this unhealthy condition.

51. C. van Vreeswijk, *Analysis of the Asynchronous State in Networks of Strongly Coupled Oscillators*, Phys. Rev. Lett. **84**, 5110–5113 (2000).
52. S. Ponten, F. Bartolomei, C. Stam, *Small-world networks and epilepsy: Graph theoretical analysis of intracerebrally recorded mesial temporal lobe seizures* Clinical Neurophysiology **118**, 918–927 (2007).
53. G. Aquino, M. Bologna, P. Grigolini, B.J. West, *Beyond the Death of Linear Response: 1/f Optimal Information Transport*, Phys. Rev. Lett. **105**, 040601 (2010).

# RENEWAL PROCESSES IN THE CRITICAL BRAIN

PAOLO ALLEGRINI,*,† PAOLO PARADISI,‡ DANILO MENICUCCI*,†
and ANGELO GEMIGNANI*,†,§

**Istituto di Fisiologia Clinica (IFC-CNR), Via Moruzzi 1, 56124 Pisa, Italy*
†*Centro EXTREME, Scuola Superiore Sant'Anna, Piazza Martiri della Libertà 7, 56127 Pisa, Italy*
‡*Istituto di Scienza e Tecnologie dell'Informazione "A. Faedo" (ISTI-CNR), Via Moruzzi 1, 56124 Pisa, Italy*
§*Dipartmento di Scienze Fisiologiche, Università di Pisa, Via San Zeno 31, 56127 Pisa, Italy*
*E-mail: allegrip@df.unipi.it*

We describe herein a multidisciplinary research, as it developes and applies concepts of the theory of complexity, in turn stemming from recent advancements of statistical physics, onto cognitive neuroscience. We discuss (define) complexity, and how the human brain is a paradigm of it. We discuss how the hypothesis of brain activity dynamically behaving as a critical system is taking momentum in literature, then we focus on a feature of critical systems (hence of the brain), which is the intermittent passage between metastable states, marked by events, locally resetting the memory, but giving rise to correlation functions with infinite correlation times. The events, extracted from multi-channel ElectroEncephaloGrams, mark (are interpreted as) a birth/death process of cooperation, namely of system elements being recruited into collective states. Finally we discuss a recently discovered form of control (in the form of a new Linear Response Theory), that allows an optimized information transmission between complex systems, named Complexity Matching.

*Keywords*: criticality; brain dynamics; phase transition; complexity.

## 1. Introduction

### 1.1. *Complexity and the brain*

A complex system is by definition composed of many elements in non-linear interaction. Neural processing is characterized by integration, each neuron collecting input from many others, and sending it to many others. The process of information spreading/reduction, triggering decisions, is of paramount complication, so it is not surprising that it is taken as a complexity paradigm. However, complexity is not a synonym of complication but also implies emergence: Many elements develop collective modes dominating the system's dynamics.

This fact is at the basis of the thermodynamic processes of phase transitions. The equilibrium properties and the non-equilibrium dynamics of collective observable may become independent of the underlying microscopic equations and, as function of some macroscopic parameter (such as temperature or energy), critical points exist marking the border between macroscopic determination and macroscopic indeterminacy. While 'below' the critical points a single thermodynamic solution exists, 'above' it multiple solutions are available. Thus, above the critical point, as the system explores only one solution, the entire collection of possible microscopic states of the alternative solutions remains unexplored. How does this happens? Approaching the critical point from below the macroscopic fluctuations around the equilibrium become larger, with more extended correlations both in space and time, till the correlation scales become infinite at the critical point. Above the point the solutions separate, and the system becomes non-ergodic. The mathematical notion of infinity means that, for example, the fluctuation autocorrelation in time (equivalently the system relaxation after a large fluctuation) is no longer an exponential decay with a finite (though becoming larger) scale, but follows a law characterized by scale invariance. This phenomenon is normally identified with $1/f$ noise.

### 1.2. *Brain activity, $1/f$ noise, and criticality*

1/f noise has been revealed in neural activities by many authors, starting from Novikov.[1] The experimental discovery of 1/f noise in the brain has normally been interpreted as an indication of self-organization. It is worth mentioning that the theory of self-organized criticality (SOC),[2] frequently invoked to understand the complex emergent output of neural activity,[3–5] has been originally introduced to explain the origin of 1/f noise.

What is the cause of these highly correlated fluctuations? Are they generated by the oscillatory neural synchronization advocated in literature[6,7] or by the spontaneous phase-transition process of many interacting neurons proposed elsewhere[8]? Maybe that the two visions are in fact related one to the other, or maybe we are in the presence of yet to be uncovered complex dynamic processes. The phase-transition conjecture, dating back to the pioneering work of Turing,[9] fits the criticality condition stressed in the neuro-physiological literature (see, e.g., Chialvo,[5] or, more recently, Werner[10]). Moreover, in the neuro-physiological literature there is increasing conviction that brain dynamics are characterized by quakes and/or avalanches.[3–5] All this literature supports the hypothesis of a critical brain, where inverse-power law correlations are present both in space and time.

According to Turing,[9] a subcritical brain would quench information, and a supercritical brain would undergo information explosion. A thinking machine has to live at the border between these macroscopic conditions. Since these seminal ideas more experimental data have been reported supporting this view, a notable example being the discovery of intermittent avalanches by Beggs and Plenz,[4] who invoke SOC,[2] which is able to explain their findings.[11] In this case the model consists in a branching process of information transport through layers of neurons, which, in terms of a parameter describing the probability of successfully transmitting a signal at each step, can be subcritical (information quenching, yielding null information transfer), critical (optimal information exchange) and supercritical (information explosion yielding saturation). This particular model predicts, at the critical point, neural avalanches whose size and durations are inverse-power-law distributed, with indexes respectively 1.5 and 2.0, in agreement with the experimental data of Beggs and Plenz.[4]

Criticality in human brain has been recently reviewed by Chialvo,[12] focusing on inverse-power laws and on the presence of avalanches, and also by the scale-free-network representation of cross correlations between pairs of units, superimposable to what can be found in a dynamical Monte-Carlo of a critical Ising model. Another illuminating review is the work of Werner,[10] showing that inverse-power laws are ubiquitous in the brain, from single cells to conciousness.

## 2. Rapid Transitions: A Serial Process Visiting Metastable States

At one of the highest levels, *i.e.* at the level of consciousness, it has been hypothesized that the dynamical pattern of "thoughts" is a serial conscious output which integrates non-conscious units which are in fact working in a highly parallel way, and that the serial, operational time is in fact characterized by the system visiting meta-stable states, with information only increasing at the transitions between a metastable state and the next.[13] This hypothesis, founded on the seminal idea of James,[14] provide a non-animistic answer to the questions on how to reconcile parallel computing with serial thinking.

This dynamical interpretation, namely the system visiting metastable states, with sporadic transitions, can be operatively associated to segmentation in electroencephalogram (EEG), namely the fact that EEG signals frequently looks like juxtapositions of epochs with fixed frequency and amplitude, with abrupt changes, called rapid transition processes (RTP), from epoch to epoch.[15]

## 2.1. *Detection of rapid transition processes*

We performed a 32-channel EEG 5 minutes' recording on thirty subjects, from 20 to 30 years old, comfortably seated in an armchair in a semi-dark and sound-attenuated room, instructed to keep the eyes closed, and avoid any structured thinking, like calculations or intentionally memories recalling. All subjects signed an informed consent according to the University of Pisa Ethical Committee guidelines. Experimental sessions were performed between 3 PM and 5 PM. Impedances of the recording electrodes were in the range of 2-5KOhm, with a sampling rate of 500Hz. All electrical potentials were reference to electrode FCz, placed slightly in front of the head vertex. EEG raw data were notch filtered at the line frequency (50Hz) and then, in order to obtain unipolar recordings, re-referenced to the instantaneous average of the signals recorded from the earlobes. Reference electrode has been excluded from the analysis, together with two electrodes, placed at the earlobes. Movement artifacts were visually detected and discarded by an EEG-expert.

We looked for the abrupt changes in EEG amplitude that glue EEG signal segments with regular amplitude and wave shape. These points (events) have been called Rapid Transition Processes (RTPs).[15] The segmentation algorithm consists of in two main stages: a first preliminary identification of the RTPs and a successive statistical selection of the actual RTPs. The first stage, is in turn composed by two steps: the estimation of the envelope (*i.e.*, of the local amplitude) of each EEG signal and the detection of the abrupt changes in the EEG local amplitude time series. In the first step we Hilbert transform[16] the EEG signal, and consider the modulus of the corresponding output. The obtained series constitutes the so called *test sequence* (*TS*).[17] In the second step we smooth the test sequence with a moving average (700msec window length) thus producing a so called *level sequence* (*LS*). In presence of rapid amplitude changes in the EEG, the *LS* will update its values with some time delay with respect to the non-smoothed $TS$. The time locations of each intersection between $TS$ and $LS$ can thus be considered as preliminary RTPs.[15]

Some of the intersections between the two series do not correspond to a discontinuity between quasi-stationary EEG segments and thus false RTPs, related to brief anomalous peaks in the test sequence, occur. According to Kaplan *et al.*,[15] anomalous peaks (either related to very tight pairs of intersections or without a steady separation between the two series) can be effectively selected out by noticing that the slope of the $TS$ around a false RTP should be less steep than that around actual RTPs and should be similar to that of any other point in the $TS$. To filter out false RTPs we make use a smoothed time derivative of the $TS$, calculated by means of a convolution with a anti-symmetrical step-like function (of window 50 msec wide, value -1 in the left part, +1 in the right, 0 outside the window). Looking at the modulus of the derivative signal, we remove segments (100 ms wide) centered around the time locations of the preliminary RTP and we estimate the probability density function of the remaining signal. On the basis of this distribution, a threshold aiming at discriminating actual from false RTP has been calculated as the 95th percentile of the distribution. Namely, we required that the $TS$ at actual RTPs exhibits high slope absolute values, that are uncommon among slopes of $TS$ generic points. A schematic representation of the segmentation procedure is shown in Fig. 1.

### 2.2. *Multichannel RTP detection*

The identification of RTP concurrent among EEG channels (MC-RTPs) allows inferences about the dynamics of large-scale cortical interactions in recruitment/selection of neuronal populations forming metastable assemblies.[19] In fact, due to the integrated functioning of the brain, RTPs happen simultaneously in two or more EEG channels more often than expected by chance and the study of waiting times between MC-RTPs allows the estimation of the lifetime of the large-scale functional assemblies of neurones. From the multivariate series of RTP locations, the time locations sequences of those RTP concurrent (simultaneous) in $N > 2$ electrodes have been constructed. Simultaneous RTPs are those events in which the time lap between two transitions is less or equal to a threshold value $\Delta t$ called "lumping time". We take $\Delta t = 2$ms, *i.e.*, the time resolution of our recordings.

A complete set of analysis aimed at evaluating the parameters (and especially $\mu$) is reviewed in the next Subsections, where we study diffusion processes driven from the waiting times, with different rules. For all the rules it is possible to evaluate how the second moment of the corresponding diffusion grows with time, and connect that index (Hurst's index $H$) with $\mu$. It is also possible to evaluate the rescaling of the spreading distribution of displacements in time (scaling index $\delta$) in terms of $\mu$. When noise is present the treatment is generalized, for the second moment analysis, to unravel the noise intensity (see Section 4), making it possible to get a robust measure of all the three indexes ($\mu$, $T$, and noise intensity).

### 3.2. *Event-driven random walks*

The rules herein described show to what an extent ideal artificial signals driven only by crucial events generate a diffusion process whose scaling properties depend on the statistics of crucial events. Now we summarize the exact theoretical predictions stemming from renewal theory, which are the basis of the methods of analysis.

Let us remind that, in general, the statistical properties of the variable $x(t)$ driven by an erratic velocity $\xi$ are related to the autocorrelation function of the signal itself via

$$C(t) \equiv \frac{\langle y(0)y(t)\rangle}{\langle y^2\rangle} = \frac{1}{2}\frac{d^2(\langle \Delta x^2(t)\rangle - \langle \Delta x(t)\rangle^2)}{dt^2}, \tag{1}$$

where $y = \xi - \langle \xi \rangle$, and $\Delta x \equiv x(t_w + t) - x(t_w)$, $t_w$ denoting the beginning of the moving windows, over which the statistics are built and averages $\langle\cdot\rangle$ are performed (time averages). The Fourier transform of $C(t)$ is the power spectrum. The function $p(\Delta x, \Delta t)$ is the probability density of finding a displacement around $\Delta x$ in a time window of length $\Delta t$. We have a scaling relation if

$$p(\Delta x, \Delta t) = F\left(\frac{\Delta x}{\Delta t^{\delta}}\right)\frac{1}{\Delta t^{\delta}}, \tag{2}$$

where $F(x)$ is a normalized distribution density. The first factor on the r.h.s. of (2) tells us that $\Delta x \sim \Delta t^{\delta}$, while the second factor is only normalization. It is useful to study the second moment of the displacement, namely the function which is differenciated twice in Eq. (1), defining an index $H$ via

$$\langle \Delta x^2(\Delta t)\rangle - \langle \Delta x(\Delta t)\rangle^2 = K\Delta t^{2H}, \tag{3}$$

where $K$ is a constant. If the scaling were perfect, as in fractional Brownian motions, $H = \delta$. When the diffusion process is driven by renewal events, $H$

and $\delta$ are slightly different in the $1/f$ regime studied herein. It is however always true that Eq. (3), making use of (1) and of a Tauberian theorem, yields $\eta = 2H - 1$ and that $H \simeq \delta$ when $\eta \simeq 1$, namely in the case of ideal $1/f$ noise. Notice further that the generated artificial signal has long-range correlations when $H \neq 0.5$.

With Rule #1, termed the Asymmetric Jump (AJ) by Grigolini *et al.*,[23] the signal is kept constant with $\xi(t) = 0$ within the laminar region, and $\xi(t) = 1$ at the time of an event occurrence. This rule generates

$$\delta = \mu - 1, \quad H = \frac{\mu}{2} \tag{4}$$

for $1 < \mu < 2$, and

$$\delta = \frac{1}{\mu - 1}, \quad H = 2 - \frac{\mu}{2} \tag{5}$$

for $2 < \mu < 3$.

With Rule #2, termed the Symmetric Velocity (SV),[23] the signal is kept constant with $\xi(t) = \pm 1$ within the laminar region, and a change of sign may or may not occur, at the time of an event occurrence, according to a coin-tossing procedure. This rule yields

$$\delta = 1, \quad H = 2 - \frac{\mu}{2} \tag{6}$$

for $1 < \mu \leq 2$, and

$$\delta = \frac{1}{\mu - 1}, \quad H = 2 - \frac{\mu}{2} \tag{7}$$

for $2 < \mu < 3$. It is important to stress that for $\mu < 2$ the second-moment prediction is apparently unphysical given the fact that in the traditional stationary case the scaling $H = 1$, the well known ballistic condition, is the impassable maximum scaling value: no diffusion can spread faster than walkers running with constant velocity with no direction change. However, if we abandon the traditional stationary perspective and we take into account the ergodicity breakdown generated by $\mu \leq 2$ we find that this apparently impassable limit can be overcome.[24]

Finally, with Rule #3, termed the Symmetric Jump (SJ),[23] the signal is kept constant with $\xi(t) = 0$ within the laminar region and $\xi(t) = \pm 1$ at the time of an event occurrence, selected by coin-tossing. This rule generates

$$\delta = H = \frac{\mu - 1}{2} \tag{8}$$

for $1 < \mu \leq 2$, and

$$\delta = H = \frac{1}{2}, \tag{9}$$

for $\mu > 2$.

The joint use of these different rules allows settling some ambiguities. For instance, we see that according to Eqs. (4) and (5) of rule # 1, (AJ) the same value of $\delta$ fitting the condition $0.5 < \delta < 1$ is compatible with both $\mu = 1 + \delta < 2$ and $\mu = 1 + 1/\delta > 2$, respectively, thereby assigning to rule # 3 (SV) or #2 (SJ) the crucial role of establishing if $\mu > 2$ or $\mu < 2$, given the fact that in the former case rule #3 would yield $\delta = (\mu - 1)/2 < 0.5$ and in the latter $\delta = 0.5$. The detection of $H$ for the signal generated using rule #1 (AJ) is affected by a similar ambiguity for the evaluation of $\mu$. The comparison of the 3 rules makes the analysis able to deal with spurious events, which can be treated as superimposed noise. Moreover, the presence of superimposed noise, may make rule # 3 useless, as the noise process yields a diffusive component with $H = \delta = 0.5$, which may wash out the subdiffusive terms for $1 \leq \mu \leq 2$, while it just sums up the already standard-diffusive terms for $2 \leq \mu \leq 3$ (again yielding $H = \delta = 0.5$). Finally, for rule #2 (SV), the evaluation of $\delta$ is numerically unstable, while for the diffusion index $H$ it happens that in the presence of noise the resulting diffusion is not trivially a linear superposition of the separate diffusions stemming form the renewal process and from the noise, (which is the case for rules #1 and #3).

### 3.3. *Diffusion Entropy and Detrended Fluctuation Analysis*

To keep the paper self consistent, here we review the methods for estimating $\delta$ and $H$, through, respectively, Diffusion Entropy (DE) and Detrended Fluctuation Analysis (DFA)

Given a diffusion process

$$\dot{x} = \xi \Rightarrow \Delta x = \int_{t_w}^{t_w + \Delta t} \xi(t')dt', \tag{10}$$

the three rules determine the value of the variable $\xi$ with respect to time $t$. For each rule we determine $p(\Delta x, \Delta t)$ and calculate the Shannon Entropy of the diffusion process (hence the name Diffusion Entropy)

$$S(\Delta t) \equiv - \int_{-\infty}^{+\infty} p(\Delta x, \Delta t) \ln p(\Delta x, \Delta t) d\Delta x. \tag{11}$$

Using the scaling condition (2), it is straightforward to prove that

$$S(\Delta t) = \delta \ln \Delta t + S_0, \tag{12}$$

where $S_0 = -\int_{-\infty}^{+\infty} F(x) \ln F(x) dx$. Notice that the scaling is in fact asymptotic, namely it is only exact for $t \to \infty$. There is however always a time $T$ denoting the time it takes for this asymptotic behavior to be reached. To take this property into account we assume

$$S(\Delta t) = \delta \ln(\Delta t + T) + S_0. \tag{13}$$

The DE method of analysis makes it possible for us to detect the scaling index $\delta$ by plotting the Shannon functional and thus determine $\delta$ by fitting the r.h.s. of Eq. (13) to the data. Since the DE technique rests on a smooth evaluation of the distribution density, rule #2, corresponding to SV, is not numerically reliable, due to data sparseness. In particular the support of $p(\Delta x, \Delta t)$ is as large as $\Delta x$ assuming all possible values in the interval $[-\Delta t, \Delta t]$, with many "holes". This is especially crucial for the DE analysis, which rests on the evaluation of the probability density function, and therefore, from a numerical standpoint, on the size of the histogram bins. On the other hand, for rules #1 (AJ) and #3 (SJ) DE rapidly converges, as the support of $\Delta x$ keeps its compactness for larger values of $\Delta t$, with respect to rule #2 (SV). In these cases the results are stable for a wide range of bin sizes. Results for Subjects 1 (red squares), 21 (green circles), 29 (blue triangles) are reported in Fig. 2. The standard-diffusive behavior of Rule #3 suggests $\mu > 2$. Different (but similar) results for our subject yield, using Rule #1, $\mu = 2.16 \pm 0.16$ for the population of our experiment.

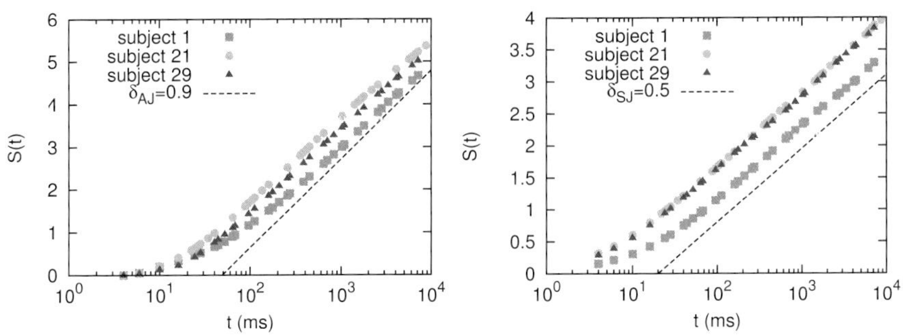


Fig. 2. Diffusion Entropy analysis for Subjects 1 (red squares), 21 (green circles), 29 (blue triangles). In the left panel Rule #1 is used and the black straight line is an eye guide for $\delta = 0.9$; In the right panel (reproduced with permission from Allegrini *et al.*[18]) Rule #3 is used and the black straight line is an eye guide for $\delta = 0.5$.

and $\psi_n^0(t) = \delta(t)$ is a Dirac $\delta$. For $t_a = \infty$ $R = 1/\langle \tau_n \rangle$,[28] so

$$\psi_n^\infty(\tau) = \frac{1}{\langle \tau_n \rangle} \int_{-\infty}^{0} \psi_n(\tau - t')dt' = \frac{\Psi_n(\tau)}{\langle \tau_n \rangle}. \tag{26}$$

Notice that before the exponential cutoff in (21) we have the joint action of two inverse-power laws, with indexes $\mu - 1$ and $\mu - 2$. If $\gamma$ were vanishingly small, the asymptotic behavior would be dictated by the index $\mu - 1$. We recall that the treatment herein holds when $\mu > 2$, but it can be generalized for $\mu < 2$, where $\Psi_n^\infty(t)$ tends to a non-vanishing constant, while $\langle \tau \rangle \to \gamma^{-1}$ as $t \to \infty$. Numerically, the presence of two inverse-power decays and the exponential cut-off can make the decay of the survival probability indistinguishable from a stretched exponential.

This is apparent in Fig. 3, showing $\Psi(t)$ for typical cascading events (RPT in more than one channel) stemming from multi-channel EEG. Here we show as an example Subject 5 of the experiment earlier discussed. Similar plots stem from all 30 subjects. For Subject 5 we show in detail how to estimate the parameters using both theory and artificial data stemming from a Monte Carlo. We show two sets of data points, one stemming from EEG data (open squares), and another with a Monte Carlo simulation of the model (open circles) with $\psi_n$ stemming from

$$\psi_n(\tau) = (\mu - 1)\frac{T^{\mu-1}}{(T + \tau)^\mu}, \tag{27}$$

where $T = 6.0$s and $\mu = 2.05$, with the Poisson-process $\gamma = 0.006$Hz, and the same length of the data analyzed (5 minutes, sampling ratio 2ms). The curves stemming from the two point sets are indistinguishable. Notably, the dotted line, indicating a best stretched-exponential fit for $\Psi(t)$ seem to accurately describe both data sets, as earlier stated. It is in fact indistinguishable from data for $t > 30$ms. We now stress that a stretched exponential behavior has been reported to be typical of non-renewal processes with long-range correlations. We remark here that also our CMM is non-renewal due to the presence of superimposed noise. However, we find this stretched exponential to be due to the joint action of Eq. (21) and finite statistics.

Let us use a more sensible correlational analysis on the CMM model using our three rules and DFA. In Rule #1 $\xi(t)$ is simply (20). As earlier stated, $\xi(t) = \xi_p(t) + \xi_n(t)$ and since the two sequences $p$ and $n$ are mutually independent, it is possible to separate their second moment contributions

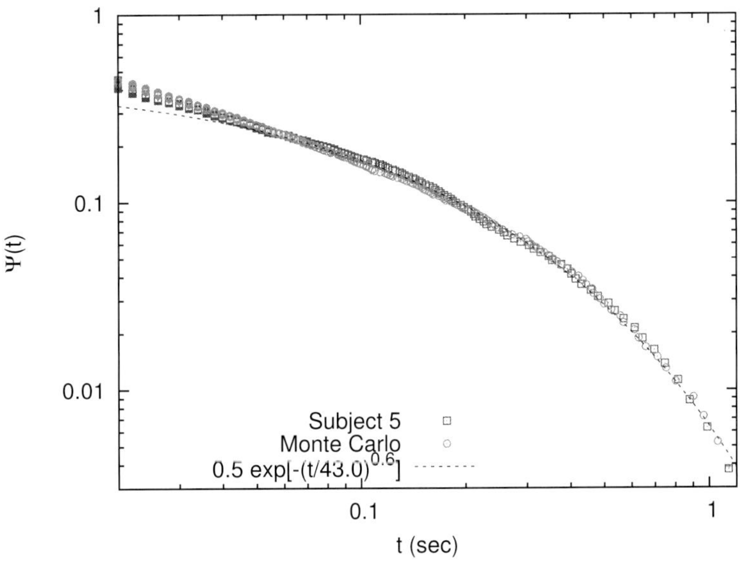


Fig. 3. Survival probability. Red open squares stem from EEG RTPs data. Green open circles stem from a numerical simulation of the CMM with $r = 0.006$ Hz, $T = 6.0$s and same statistics of the data. Blue dotted line is a stretched exponential $A\exp[(t/B)^{\alpha}]$, with $A = 0.5$, $B = 43$s, $\alpha = 0.6$.

$x_p$ and $x_n$ and we have that

$$\sigma^2(t) = \langle x_p^2(t)\rangle_p + \langle x_n^2(t)\rangle_n, \tag{28}$$

where the suffix $p$ ($n$) on the angular brackets means the average on the $p$ ($n$) sequence alone. The Poisson process yields standard diffusion $\langle x_p^2\rangle_p \propto t$, while for renewal non-Poisson processes $\langle x_n^2(t)\rangle_n \propto t^{4-\mu}$. The non-Poisson process dominates in the asymptotic limit. However, the presence of the Poisson Process yields a transient $H = 0.5$ in the short-time limit. This behavior is illustrated in Fig. 4, central curves, where we compare the analyses same EEG data of Fig. 2 with the thoeretical prediction, with $\mu = 2.05$.

For Rule #3 we see standard diffusion, suggesting $\mu \geq 2$. In this case both the $n$ and the $p$ sequences yield $H = 0.5$ and, again, Eq. (28) holds true. The lower curves of Fig. 3 show this behavior nicely.

Finally, Rule #2 consists of making a walker assume a constant velocity $\pm 1$ within laminar regions, chosen with a coin-tossing procedure for every region. In this case (28) does not hold: While for this rule we again have $\langle x_p^2\rangle_p \propto t$ and $\langle x_n^2(t)\rangle_n \propto t^{4-\mu}$, the asymptotics are dominated by the Poisson process, while the non-Poisson scaling is confined in the short-time

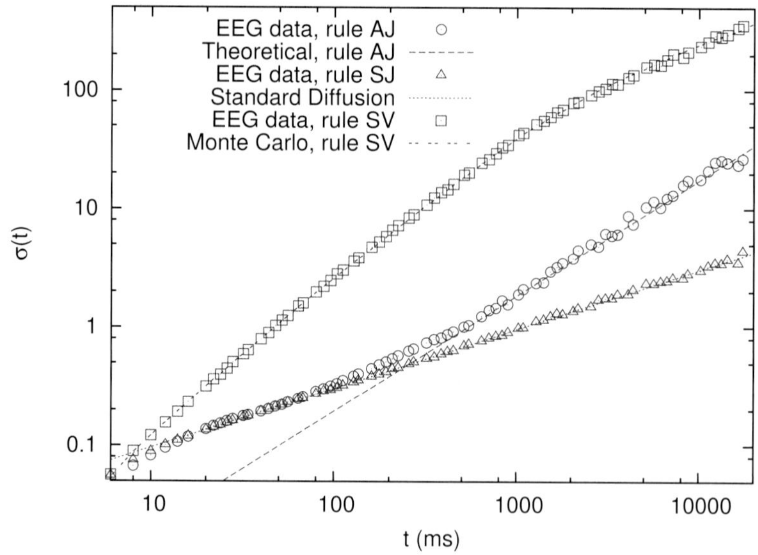


Fig. 4. DFA Analysis: Rule #1 (AJ) for EEG data (red open circles) compared with theoretical prediction $\sigma(t) \propto t^{2-\mu/2}$ (green dashed line); Rule #2 (SV) for EEG data (cyan open squares) compared with same rule on the same Monte Carlo simulation of Fig. 1 (black double-dashed line). Rule #3 for EEG data (blue open triangles) compared with an eye-guide describing standard deviation ($H = 0.5$, violet dotted line).

limit. Fig. 3 (upper curves) illustrates this behavior both for the data and the Monte-Carlo model.

## 5. Complexity Matching

The complexity matching is a recently developed theory describing the transport of information from one complex system to the other. The requirement is that these systems need to be described (or driven) by a non-Poisson renewal process. A comprehensive overview has been recently published.[29] It states that the response of complex systems to periodic or Poisson stimuli quenches off, but it does not in the presence of stimuli with the same order of complexity. Numerical evidence of the quenching of harmonic stimuli has been provided using Monte Carlo models,[30] while it was experimentally proved.[31] A recent paper makes the intriguing conjecture that this is at the basis of habituation.[32] Although this behavior seems to be of some importance, as it allows a filter for periodic components, more interesting is the non-vanishing response to complex stimuli. While the theoretical framework is still under theoretical development, some rigorous results have recently appeared.[33] The theory resembles and extends the condition of aperiodic stochastic resonance. Here the resonance condition refers to matching the complexity indexes, with optimal response at the 1/f noise condition. The theory is based to an event-driven Linear Response Theory.[34] An easy-to-follow derivation was recently reported.[35]

It is worth emphasizing that complexity matching, at least inferentially, can be invoked to explain some unexpected coincidences, namely that the indexes of resting-wakefulness EEG,[22] of human language[26] and of heart-rate variability[36] match. In particular, the accurate analysis based on salient words, performed by Allegrini *et al.*, yielded $\mu_{language} \approx 2.1$ for human language. Moreover, an analysis similar to what was described herein was performed on heart-rate variability,[36] resulting in evaluating both $\delta$ and the noise intensity for a CMM model. In this case it was calculated a mean $\delta$ for healthy individual, which, according to the theory,[23] corresponds to $\mu_{heart} \approx 2.20$. It is remarkable that these indexes $\mu$ (for the brain activity), $\mu_{language}$ and $\mu_{heart}$ are not only similar and belonging to interval $[2, 3]$, but they are also near the inferior limit of this interval, where the most recent theories[33] predict a maximal information transfer. Were this efficient flow of information confirmed by suitable experiments, involving both central and autonomic activity recordings, with concurrent tasks of speech or writing production, this would open the door to a 'holistic' fractal physiology having its bases on 'axes', like the heart-brain axis.

We have stressed that the complexity matching rests on renewal processes, and we have mentioned how critical systems output renewal signals. We end the Chapter by stressing that it is this very renewal character that makes it possible to have a serial process of entropy production. Is this process "thinking" or "consciousness"? We don't know at this time. It is apparent that some part of our cognitive experience passes through language, which is in fact serial and renewal. Here and in previous papers[22,37] we have proved the presence of a renewal component on global changes in EEGs. Are we measuring something related to consciousness? In another paper[18] we left the question open, as we reported results on the frequency over different electrodes of being recruited in global RTPs occurrences. The result, on the one hand, resembled those of Beggs and Plenz,[3] with an inverse-power law distribution of the number of electrodes involved at each time. On the other hand, the more involved electrodes seemed to overlap with a default-mode network,[38] which is thought to be involved in self-recognition tasks, but is also active in states of unconsciousness.

At the time we are writing, preliminary results, that will be presented elsewhere, show that the spatial structure is conserved both during wakefulness and sleep, while the renewal process, yielding $\mu \sim 2$ breaks down during deep sleep, and recovers back during Rapid-eye-movements sleep, namely during dreaming.

## References

1. E. Novikov, A. Novikov, D. Shannahoff-Khalsa, B. Schwartz, J. Wright, *Phys. Rev. E* **56**, R2387 (1997).
2. P. Bak, C. Tang, K. Wiesenfeld, *Phys. Rev. Lett.* **59**, 381 (1987).
3. D. Plenz and T.C. Thiagarjan, *Trends Neurosci.* **30**, 101 (2007).
4. J.M. Beggs, D. Plenz, *J. Neurosci.* **23**, 11167 (2003).
5. D.R. Chialvo, *New Ideas Psychol.* **26**, 158 (2008).
6. J.M. Medina, *Phys. Rev. E* **79**, 011902 (1-8) (2009).
7. M.G. Kizbichler, M.L. Smith, S.R. Christensen, E. Bullmore, *Plos Computational Biology* **5**, 1 (2009).
8. S, Bianco, E. Geneston, P. Grigolini and M. Ignaccolo, *Physica A* **387**, 1387 (2008).
9. A.M. Turing, *Mind* **59**, 236 (1957).
10. G. Werner, *Front. Physiol.* **1**, 15 (2010).
11. L. De Arcangelis, C. Perrone-Capano, H. J. Herrmann, *Phys. Rev. Lett.* **96**, 028107 (2006).
12. D.R. Chialvo, *Nat. Phys.* **6**, 744 (2010).
13. A.A. Fingelkurts, C.F.H. Neves, *Phys. Life Rev.* **7**, 195 (2010).
14. W. James, *The Principles of Psychology*, Vol. 1. (Dover, New York, 1890).
15. A. Ya. Kaplan, A.A. Fingelkurts, A.A. Fingelkurts, S.V. Borisov, B.S. Darkhovsky, *Signal Process.* **85**, 2190 (2005).
16. N.E. Huang *et al.*, *Proc. Roy. Soc. Lond.* **454**, 903–993 (1998).
17. B.E. Brodsky, B.S. Darkhovsky, A. Ya. Kaplan, S. L. Shishkin, *Comput. Meth. Programs Biomed.* **60**, 93 (1999).
18. P. Allegrini, P. Paradisi, D. Menicucci and A. Gemignani, *Front. Physiol.* **1**, 128 (2010).
19. A.A Fingelkurts *et al.*, *Hum. Brain Mapp.* **22**, 27 (2004).
20. Y.F. Contoyiannis, F.K. Diakonos, *Phys. Lett. A* **268**, 286 (2000).
21. L. Silvestri, L. Fronzoni, P. Grigolini, P. Allegrini, *Phys. Rev. Lett.* **102**, 014502 (2009).
22. P. Allegrini *et al.*, *Phys. Rev. E* **80**, 061914 (2009).
23. P. Grigolini, L. Palatella, G. Raffaelli, *Fractals* **9**, 439–449 (2001).
24. A.K. Kalashyan, M. Buiatti, P. Grigolini, *Chaos, Solitons & Fractals*, **39**, 895 (2009).
25. C.-K. Peng *et al. Phys. Rev. E* **49**, 1685 (1994).
26. P. Allegrini, P. Grigolini, L. Palatella, *Chaos, Solitons and Fractals* **20**, 95–105 (2004).

27. P. Allegrini, M. Barbi, P. Grigolini, B.J. West, *Phys. Rev. E* **52**, 5281 (1995); P. Allegrini, P. Grigolini, B.J. West, *Phys. Lett. A* **211**, 217 (1996).
28. P. Allegrini, G. Aquino, P. Grigolini, L. Palatella and A. Rosa, *Phys. Rev. E* **68**, 056123 (2003)
29. B.J. West, E.L. Geneston, P. Grigolini, *Phys. Rep.* **468**, 1 (2008).
30. P. Allegrini, M. Bologna, P. Grigolini, B.J. West, arXiv:cond-mat/0612303 (2006).
31. P. Allegrini, M. Bologna, L. Fronzoni, P. Grigolini, L. Silvestri, *Phys. Rev. Lett.* **103**, 030602 (2009).
32. B.J. West and P. Grigolini, *Physica A* **389**, 5706 (2010).
33. G. Aquino, M. Bologna, P. Grigolini, B.J. West, *Phys. Rev. Lett.* **105**, 040601 (2010).
34. P. Allegrini, M. Bologna, P. Grigolini, B.J. West, *Phys. Rev. Lett.* **99**, 010603 (2007).
35. P. Allegrini *et al.*, to appear in *J. Phys. Conf. Series*
36. P. Allegrini *et al.*, *Phys. Rev. E* **67**, 062901 (2003).
37. P. Allegrini, D. Menicucci, R. Bedini, A. Gemignani, P. Paradisi *et al.*, *Phys. Rev. E* **82**, 015103 (2010).
38. M.E. Raichle, A.M. Mac Leod, A.Z. Snyder, W.J. Powers, D.A. Gusnard, G.L. Shulman, *Proc. Natl. Acad. Sci. USA* **98**, 676 (2001).

energy-dominated process in the upper panel. In the lower panel a network with high energy but low information is depicted, which is unlike the thermodynamic situation. In this latter case the network with the greater information can influence the network with the lesser information, even though the latter may have the greater energy. This is an information-dominated process.

The transfer of influence from a complex network high in information to one low in information is the first attempt at a universal principle of network science and represents what we call Wiener's Rule. Of course, Wiener was speculating, giving the audience the benefit of his experience. That insight was not to be vindicated for another half century [3] and required the generalization of a number of concepts from statistical physics [4,5].

One measure of the information content of a network is provided by the probability density most often used in the determination of the negative entropy of Shannon and Wiener. An apparently ubiquitous distribution in the description of empirical complex networks is the hyperbolic, which denotes the survival probability as

$$\Psi(t) = \frac{T^{\mu-1}}{\left(T+t\right)^{\mu-1}} \tag{1}$$

that asymptotically becomes an inverse power law in time. We could list a hundred natural, social and medical phenomena that display this behavior but a few exemplars should suffice, including blackout intervals on the North American power grid, physiologic processes such as breathing and heart beats, solar flare eruptions on the sun and the time intervals between earthquakes of a given magnitude [1]. The average time between events in all the complex webs mentioned, can be determined using the probability density $\psi(t) = -d\Psi(t)/dt$, is

$$\langle t \rangle = \int_0^\infty t\psi(t)dt = \begin{cases} \dfrac{T}{\mu-2}; & \mu > 2 \text{ ergodic} \\ \infty\,; & \mu < 2 \text{ non-ergodic} \end{cases} \tag{2}$$

It is interesting that when the power-law index is in the interval $2 < \mu < 3$ the distribution has a finite first moment and the statistics are ergodic, meaning that the time average and ensemble averages yield the same result. However when the power-law index is $\mu < 2$ there are no finite integer moments and the two kinds of averages are not the same, that is, the process is non-ergodic. We shall have more to say about non-ergodicity subsequently.

The mathematical/computational verification of Wiener's insight required over half a century and uses the properties of complex networks [3]. However, in this talk we are interested in making the first applications of Wiener's Rule and not in reiterating the underlying mathematics. These details are presented in a number of the other talks in this workshop by the students who developed them. One measure of the information transfer between two complex networks is the cross-correlation between a complex network P and a complex network S being perturbed by P with $\varepsilon$ the strength of the perturbation. For our purposes it is sufficient to apply the generalized linear response theory [4,5] we previously developed to the cross-correlation function [3]:

$$C(t) = \varepsilon \int_0^t R_x(t')\Psi_s(t-t')\Psi_p(t,t')dt' \cdot \tag{3}$$

The perturbing complex network P is characterized by the non-stationary autocorrelation function $\Psi_p(t,t')$, which depends separately on the time of the last perturbation $t'$ and time of the measurement $t$. The function $R_s(t')$ is the rate of generating perturbing events at time $t'$ within the network being perturbed and is based on renewal theory [1]. The perturbed network S is characterized by the stationary autocorrelation function $\Psi_s(t-t')$, which depends only on the difference in times from the last perturbation to the measurement.

In Figure 2 the asymptotic cross-correlation function normalized to the strength of the perturbation is graphed as a function of the power-law indices of the two networks to form the cross-correlation cube. This cube displays a number of remarkable properties: 1) when the power-law indices are both equal to two there is an abrupt jump from zero correlation to perfect consensus; 2) the upper plateau indicates that when P is non-ergodic $1 < \mu_p < 2$ and S is ergodic $2 < \mu_s < 3$ there is an information response in which the perturbed network tracks the perturbing network exactly and the information transfer is maximal; 3) when P is ergodic $2 < \mu_p < 3$ and S is non-ergodic $1 < \mu_s < 2$ there is no response asymptotically and the information transfer is minimal. How a complex network responds to a perturbation by another complex network is determined by whether or not there is a mismatch in the complexity of the fluctuations in the two networks.

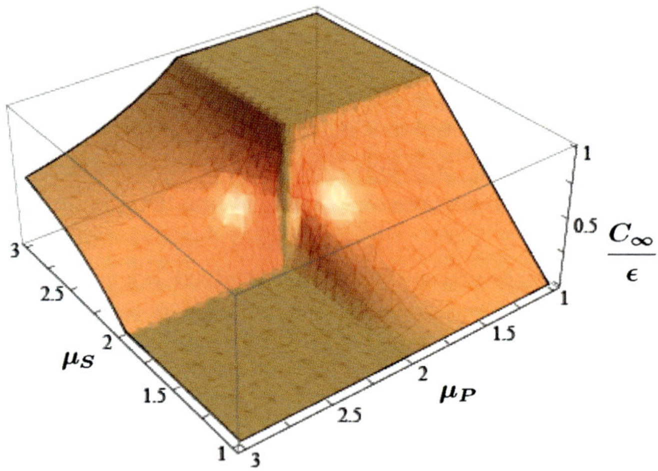


Fig. 2. The cross-correlation cube. The asymptotic cross-correlation function defined by (3) is graphed as a function of the two power-law indices of the perturbed network S and the perturbing network P. Adapted from [3].

Wiener's Rule describes the influence of the perturbing network outside the lower plateau region of the cross-correlation cube. In all regions except this one the weak perturbation significantly modifies the properties of the perturbed complex network. In the upper plateau region the perturbation by network P actually dominates the properties of the perturbed network S and reorganizes it, just as Wiener predicted. The Principle of Complexity Management (PCM) embodied in the cross-correlation cube therefore subsumes Wiener's Law.

### 3. From the Cerebral to the Celestial

Let us now turn our attention from the general formalism to some applications of the cross-correlation cube and PCM. We give two examples that probably could not be farther apart in scale and substance: How the human brain works during the process of habituation and how the Earth responds to heating by the Sun. The latter is an application of Wiener's Rule the former is not, but they are both described by the cross-correlation cube, as we shall see.

Habituation is a ubiquitous and extremely simple form of learning through which animals, including humans; learn to disregard stimuli that are no longer novel, thereby allowing them to attend to new stimuli [6]. A repetitious stimulus of unvarying amplitude and frequency content induces a response that fades over time since no new information is being presented to the animal. The lack of new information allows the brain to shift its focus from the more to the less familiar, the latter providing new information that may have survival value. However, we know the brain does not habituate to all external stimuli so let us consider two distinct kinds of stimulation; one ergodic $2 \leq \mu_p \leq 3$ and another non-ergodic $1 \leq \mu_p \leq 2$. The perturbation in the ergodic regime can be expressed as a decomposition of linear harmonic terms, which allows us to generalize a previously established result for a periodic stimulation of a complex network [7]. The non-ergodic stimulus can be drawn from a number of sources; here we choose the sequence of splashes from a dripping faucet [8] and classical music [9].

Consider first the case of highway noise coming in through the window as you lay in bed at night. This noise is typically a broad band uncorrelated random process and consequently it is ergodic. Most people habituate to this noise, meaning that the neurons no longer fire in response to this excitation, and soon fall asleep. In the present context it is possible to prove that the response of the brain to the noise fades as an inverse power law in the following way [10]:

$$D(t) \propto \frac{1}{t^{|2-\mu_s|}}. \tag{4}$$

The lower plateau in Figure 2 is the parameter region $2 \leq \mu_p \leq 3$, $1 \leq \mu_s \leq 2$ where the dynamics of the brain and the external stimulus are asymptotically statistically independent of one another. Brain activity in the non-ergodic regime $1 \leq \mu_s \leq 2$ is asymptotically unresponsive to ergodic and/or periodic stimuli; the complexity of the neuron network

11. B.J. West and W. Deering, *The Lure of Modern Science: Fractal Thinking*, World Scientific, Singapore (1995).
12. Y. Yu, R. Romero and T.S. Lee, "Preference of Sensory Neural Codifying for 1/f signals", *Phys. Rev. Lett.* **94**, 108103 (2005).
13. N. Scafetta and B.J. West, "Solar Flare Intermittency and the Earth's Temperature Anomalies", *Phys. Rev. Lett.* **90**, 248701-1 (2003).
14. B.J. West and P. Grigolini, "Sun-climate complexity linking", *Phys. Rev. Lett.* **100**, 088501 (2008).
15. B.J. West and N. Scafetta, *Disrupted Networks: from physics to climate change*, World Scientific, New Jersey (2010).
16. M. Turalska, M. Lukovic, B.J. West and P. Grigolini, "Complexity and synchronization", *Phys. Rev. E* **80**, 021110 (2009).
17. B.J. West and P. Grigolini, *Complex Webs: Anticipating the Improbable*, Cambridge University Press, New York (2010).